"*Cooperate to Compete* has the potential to awaken our generation of managers to the fact that globalization, agility and virtual organizations have arrived. Challenging our management insight, this well written book lays out the issues and opportunities; then follows up with understandable examples, summary points and 'things to think about.' The result is an informative guide we can use to thrive in this dynamic new business environment. My compliments to the authors."

—Cal J. Kirby, Corporate Vice President
Hughes Electronics Corporation

"This book contains the keys to unlocking the profitability of the firm's assets and is a must for managers desiring success in the dynamic environment of the next decades. It provides a model for understanding how companies must relate to their customers, suppliers and employees as well as to the methodologies to formulate action plans and measure their performance."

—Thomas F. Kirk, Vice President &
Chief Financial Officer
Quaker Chemical Corporation

"In *Cooperate to Compete*, the authors continue to evolve the concept of agility. Their action oriented emphasis is on the opportunities available to those who exploit the dynamic restructuring of today's markets and industries. Using an extensive collection of examples drawn from a variety of industries and services, the authors offer useful insights and advocate a workable dynamic business model. This book can help you mobilize to succeed in your market..."

—David Lando, Vice President for Engineering
& Environmental Technologies
Lucent Technologies, Bell Laboratories

"*Cooperate to Compete* makes a compelling case for the 'interprise' as the business model for success in today's marketplace. The 'interprise,' as a time based, virtual organization, teams a company's suppliers and customers into a cooperative force using the concepts of agility to bring a new dimension to business competition."

—Jack E. Swindle, Senior Vice President of Corporate Staff
Texas Instruments Incorporated

"The authors do an outstanding job of showing how utilization of worldwide communication and knowledge transfer is changing enterprises to dynamic, interactive 'interprises'."

—Dr. Hans-Jürgen Warnecke, President
Fraunhofer-Gesellschaft

Cooperate
to
COMPETE

Cooperate
to
COMPETE

Building Agile Business Relationships

KENNETH PREISS

STEVEN L. GOLDMAN

ROGER N. NAGEL

VAN NOSTRAND REINHOLD

I(T)P™ A Division of International Thomson Publishing Inc.

New York • Albany • Bonn • Boston • Detroit • London • Madrid • Melbourne
Mexico City • Paris • San Francisco • Singapore • Tokyo • Toronto

The ideas presented in this book are generic and strategic. Their specific application
to a particular company must be the responsibility of the management of that company, based on management's understanding of their company's procedures, culture,
resources, and competitive situation.

Printed in the United States of America.

Van Nostrand Reinhold
115 Fifth Avenue
New York, NY 10003

International Thomson Publishing Germany
Königswinterer Str. 418
53227 Bonn
Germany

International Thomson Publishing
Berkshire House, 168-173
High Holborn
London WC1V 7AA
England

International Thomson Publishing Asia
221 Henderson Building #05-10
Singapore 0315

Thomas Nelson Australia
102 Dodds Street
South Melbourne 3205
Victoria, Australia

International Thomson Publishing Japan
Kyowa Building, 3F
2-2-1 Hirakawacho
Chiyoda-Ku, Tokyo 102
Japan

Nelson Canada
1120 Birchmount Road
Scarborough, Ontario
M1K 5G4, Canada

2 3 4 5 6 7 8 9 10 QUEFF 01 00 99 98 97 96

Library of Congress Cataloging-in-Publication Data available upon request.

ISBN 0-442-02253-0

HD41
.P74
1996

Production: Jo-Ann Campbell • mle design • 562 Milford Point Rd., Milford, CT 06460

To Miriam, Michal and Yonat.

K. P.

To my colleagues at the Agility Forum and to the memory of Texas Instruments' late Chairman, President and CEO, Jerry Junkins, whose pioneering leadership was instrumental in the launch of the Forum.

S. L. G.

To my family, Arlene, Bruce and Deborah, and to our parents Hyman and Isabel Nagel, and Miriam and Irving Fluxgold for their support, faith, and encouragement.

R. N. N.

CONTENTS

Preface   ix
Acknowledgments   xiii

PART 1   CONNECTING WITH CUSTOMERS   1

1   Why The Interprise?   3
2   The Customer Connection   11
3   The Interprise Payoff   21
4   Dynamic Relationships   33
5   Adding Value By Subtracting Time   41
6   Products As Platforms   53
7   To Each His Own   65
8   Divide To Conquer   75
9   Global Connections   85

PART 2   MUTUALLY PROFITABLE RELATIONSHIPS   99

10   Multiple Points Of View   101
11   Balancing The Relationships   111
12   Enriching Customers   121
13   Your Customer's Customer   131
14   Price Follows Value   139
15   Building Linkages   149

| 16 | Virtual Relationships | 157 |
| 17 | The Trust Factor | 169 |

PART 3	**FROM ENTERPRISE TO INTERPRISE**	**177**
18	Teams Not Committees	179
19	Follow The Money	191
20	The People Weapon	215
21	Leading The Way	225
22	Creating The Interprise	237
23	Making It Work	251
24	Mastering Change	263
25	The Measure Of Success	275

	References and Suggested Readings	301
	Index	305
	About the Authors	311

PREFACE

*C*ooperate to Compete is about the competitive advantage that businesses are gaining as they move from arm's length to interactive relationships. The new enterprise operates with fuzzy boundaries. Its processes, systems, operations and personnel are interactively and opportunistically linked to those of customers, suppliers and partners, even competitors. We call this new interactive organization the interprise. Its competitive advantage lies in creating value by leveraging relationships to the mutual benefit of all participants.

In our earlier book, *Agile Competitors and Virtual Organizations*, we presented a model of post-mass-production, so-called "agile" commerce. We identified the distinctive market forces driving business process change today, offered a strategic model of agility-based competition and described the large-scale features of agile enterprises-in-the-making. *Cooperate to Compete* describes the interprise, which is the heart of agile competition, the means by which companies can transform themselves into agile competitors. Understanding how to make the transition from a traditional enterprise to an agile interprise effectively is the single greatest challenge facing management today, and its greatest opportunity.

Since *Agile Competitors* went on sale interest in agility as a systematic response to fundamental changes in global commerce has grown dramatically. It has spread from manufacturing to service companies, from industry to government agencies and educational institutions, and from the U. S. to western Europe, South America, India, and eastern Asia. What began as acceptance of a strategic model of agile competition has evolved into a demand for assistance in translating that model into specific and concrete action. *Cooperate to Compete* is a response to that demand. Our goal in this book is to move a giant step closer to answering the tactical question, 'You may be right about all this, but what do I do differently on Monday morning?'

In our previous book we argued that there is no one-solution-fits-all response to the challenge of gaining and sustaining competitive advantage in the agile commercial environment. The proliferation of smaller and smaller niche markets, for individualized services and products with shrinking market life-times, means that each company needs to determine for itself how it can exploit those market opportunities that match its own distinctive competencies. Each company must determine the forms of organizational change most appropriate to its market-driven operations, the optimal utilization—in-house and through alliances—of the human and technological resources needed to develop profitable customer-driven solutions-products, and the most effective application to marketing and product development of rapidly changing information and knowledge.

Cooperate to Compete prepares readers to make those determinations on behalf of their respective companies by showing: why enterprises are becoming interprises, how interprises function, and how they are being created. Numerous examples are given throughout the book of leading companies that have become successful interprises. By constantly linking operational details to an overall view, we show the relevance of each detail to the new strategic goals of competitive businesses.

Part 1 describes how the interprise goes beyond cost, quality, and timeliness of a product or service, to becoming a supporter of the customer's business or lifestyle processes. The customer comes first because without profit-generating customers any business is doomed. The point seems too obvious to make, yet companies continue to invest huge sums of money reengineering processes independent of the value to customers of the anticipated "improvements." Customers don't buy a supplier's processes, however elegant and efficient they may be. They buy products and information that provide value for them. Interactive relationships with customers allow a company to get beyond the pursuit of cost-saving to growth through the creation of new, more valuable, products and services.

Part 2 addresses techniques for leveraging inter-enterprise relationships to the mutual benefit of all participants. These include the sharing of risk, reward and information; communication; the creation of value-creating interdependencies; and trust-based inter-enterprise relationships. It describes as well the "virtual organization"model for opportunistically integrating competencies distributed among a group of cooperating companies into a single powerful "virtual" competitor.

Part 3 offers a road map for moving from the old enterprise to the new interprise, from isolation to cooperation. It describes the new responsibilities of management, the new organizational structures and the new financial metrics which are the basis of business success. We end with an exercise that companies have used successfully to plan a prioritized series of actions for changing from an arm's length enterprise to the dynamic, interactive competitor we call the interprise.

We take an active interest in the subject of contemporary competitiveness. Interaction with those readers willing to share and compare their experiences in this subject with us and each other, continues to enrich all of us. We welcome comments from readers. (e-mail address is compete@lehigh.edu)

K. P.
S. L. G.
R. N. N.

ACKNOWLEDGMENTS

Since the publication of *Agile Competitors and Virtual Organizations* in 1995, we have worked with hundreds of dedicated and capable people in companies large and small, from a broad variety of industries, universities and government agencies. Our common goal was to convert good ideas into successful practices. We have learned more than we can acknowledge from these relationships, and thank those individuals and their organizations for the learning opportunities they have provided us.

Working with Lee Iacocca and the people of the Iacocca Institute at Lehigh University, and its Agility Forum under the leadership of the industry president and CEO, Rusty Patterson, we feel privileged to be at the center of thinking about agility.

We thank the Iacocca Institute Advisory Board chairman Lee Iacocca and members Curtis Barnette, Jan Berninger, George Fisher, Douglas Fraser, William Hecht, William Hittinger, Russell Leslie, Peter Likins, Aris Melissaratos, Thomas Murrin, John Puth, David Roderick, Robert Woodson, Sr. and Jerome York, for providing guidance, advice and feedback.

We thank the Agility Forum Board, chairman Aris Melissaratos and members Rhonda Gross, Bruce Haines, Cal Kirby, Peter Likins, John Puth and Don Runkle for continuing to provide ideas, inspiration and support.

The members of the Agility Forum's Strategic Analysis Working Group have provided particularly valuable assistance, especially Alec Lengyel, Ted Goranson, Norm Kuchar, Mike McGrath, Bill McNally, John Mills, Joe Off, Tom Shaw, Dan Shunk, and Dale Snyder. These people were largely responsible for organizing the generic model used in chapters 22–25.

The members and leaders of the Agile Learning Center at Lehigh University, Donald Bolle, Mike Bolton, Letitia Conkey, Napoleon Devia, Lin Erickson, Vinay Govande, Bob Gustafson, Jennifer Montemurro, Gaurav Mehrotra, Christine Pense, Ron Sedlack, Joan Smith, Sekar Sundararajan, George Wood, and Emory Zimmers, Jr. have contributed valuable suggestions.

We appreciate the advice and experience of Ted Nickel, past president and Bill Adams, president of the Agile Web group of companies. We also acknowledge the contributions of the Agile Web leaders: Andrew Behler, Ray Biery, Jeffrey Bilger, Anthony Godonis, Charles Hagan, David Krisovitch, Jeffrey Lamm, Thomas Martin, Jeffrey McGinley, Terrance McGinn, Thomas Panzarella, Jack Pfunder, Joseph Rado, Earl Ruckdeschel, Michael Stirr, William Straccia, Stewart White, James Williams, and Steve Yohe.

The staff of the Agility Forum, who have supported our efforts and with whom it is a pleasure to be working and collaborating, are: Pearl Anderson, Kathy Bailey, Joyce Barker, Vicki Bawden, Dawn Bold, Dan Church, James Davis, John Dieser, Tom Falteich, Marc Field, Devon Frey, Bob Gilbert, Carol Graham, Stephen Heacock, Pat Heimbach, Jane Hontz, Lis Hoveland, Mike Kirkman, Dave Knies, Lynn Landrock, LeeAnne Leckey, Bill Magerman, Tammy Mallen, Marcia Martin, Elizabeth Massa, Judy Mattei, Jenny Nepon, Sandi Odrey, Rob Opitz, Pamela Perkins, Al Philpotts, Rose Reinhard, Gary Sadavage, Jason Salgado, Karen Serulneck, Troy Silfies, Marie Ann Tassos, Susan Tropeano, Beverly Ward, and Kristen Wecht.

The continued, proactive support of the following people and organizations has been of special value to us in developing this book: Juval Bar-On, Harry Dickman, Henry Duignan, Robert Hall, Jim Hughes, Ron Hysom, Fred Kovac, David Lando, Larry Leson, Tom Neilssen, Alan Pense, Manash Ray, Tom Ries, Phil Roether, Sal Scaringella, Offer Shai, Dan Stern, Jim Sims, Les Tuerk, Gary Thompson, Fern Webber and Bud Wyble, Jane Carpenter, Emily Gohn and all the Wharton Advanced Management Program (AMP) students, Nick Franklin and his Agility Implementation Team at Delphi Saginaw.

The support and encouragement of the president of Ben Gurion University of the Negev, Avishay Braverman, and the rector, Nahum Finger, is appreciated.

We appreciate the special efforts of Gene Mater in preparing a number of illustrations. John Mies of Mack Trucks, Inc. and Mark Greiner and Howard Sutton of Steelcase kindly made available graphic material. Ron Shulman provided Figure 19.6, and Nick Franklin of Delphi Saginaw Steering Systems made available Figure 4.3. Figure 19.5 is reproduced from page 67 of The Haystack Syndrome (see references), by permission of North River Press.

Examples of company behavior reported throughout the book were compiled primarily from personal interviews, and secondarily from the business media, especially the *Wall Street Journal*, *New York Times*, *Financial Times*, *Business Week*, *Fortune*, *Forbes*, and *Wired*.

Our work has been ably supported at our publisher by Marianne Russell, John Boyd, Chris Bates, Tom Cardemone, Jon Herder, Jacqueline Jeng, and especially the talented and patient editor, Rosemary Ford.

Cooperate
to
COMPETE

Part 1

CONNECTING WITH CUSTOMERS

CHAPTER 1

WHY THE INTERPRISE?

B usinesses are undergoing as profound a change today as they did at the end of the nineteenth century, when the modern industrial corporation was invented.

The drivers of such profound business process change include:

- The worldwide spread of education and technology, leading to intense and increasingly global competition and accelerating rates of marketplace change.
- The continuing fragmentation of mass markets into niche markets.
- More demanding customers with higher expectations.
- The spread of collaborative production with suppliers and customers who comprise the value-adding chain.
- The increasing impact of changing societal values, such as environmental considerations or job creation, on corporate decision making.

The old fashioned business environment, in which each company could be managed in isolation, has rapidly changed into one in which decisions made by one business directly impact decisions in other businesses. Management today involves continuous interactivity between businesses.

Some businesses are successfully adapting to this new, more dynamic environment. They can readily respond to the rapidly changing demands of their customers and the marketplace. They strive to understand and meet the needs of their customers. They provide more than "good service," they become part of their customers' businesses. They forge strong, enduring bonds with suppliers to enhance mutual goals. The internal organization of these companies encourages an adaptive entrepreneurial attitude among staff who recognize that the company's success is tied to their ability to support their clients. They are as interactive and international as is the new culture of the internet. We call this new kind of organization an interprise.

> **An interprise exhibits increased integration with the business processes of customers, more cooperation with suppliers, and an entrepreneurial internal environment.**

THE OLD ORGANIZATION CHART

This organization chart, still used by many companies, divides workers by department. In a hierarchically controlled organization, only the president or owner is aware of the general picture. Each manager knows only what is relevant for his or her department, and each worker knows only his or her specialized job. This style of management is gradually being eradicated and replaced by process-focused management. As layers of management are removed and structures changed, the old organization chart can no longer summarize the essential processes and connections in a modern company. The need to deal with incessant change creates new capabilities, such as aligning with customers and suppliers, and the transfer of knowledge along with a physical product. The old corporate pyramid has been replaced by an entrepreneurial internal and extended structure, where each employee is at the center of an interactive web.

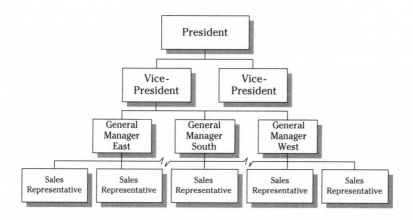

Figure 1.1 The Corporate Pyramid.

THE INTERPRISE CHART

Figure 1.2 The Process Chart.

The process chart shown above illustrates key activities in a modern company. It is based upon a diagram often used by planners to define an interactive system, and it illustrates how a business operates as a process. In this process a business makes money by taking inputs, shown at the left, and converting them to outputs, shown at the right, subject to the limitations and conditions shown at the top of the diagram.

Among these limitations and conditions are some a manager can control, and some that he cannot. The unchangeable external conditions include currency fluctuations, economic changes, laws, and the behavior of competitors (although a good company can inflict a changing competitive environment on its competitors, which we discuss later in the book). Management, together with all the people in an interprise, fosters an interactive relationship between the company and the customers who pay their way—as well as its supporting "food chain" of value-adding suppliers—and develops new internal structures and relationships within the company.

The most important new items in business processes, that we have observed in many modern successful interprises and are referenced in many studies, are shown in Figure 1.3. These are the key, specific factors that management faces in today's environment:

- Inputs: multicompany cooperating resources, including total value chain management and supplier integration
- Internal structure: an adaptable, entrepreneurial, knowledge-based learning organization
- Outputs: cooperation with customers to provide individualized, total solutions
- External influences: a world of relentless, accelerating change

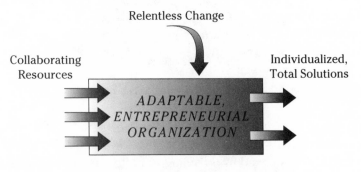

Figure 1.3 The Interprise Chart.

In the opening of his second book on reengineering, Champy writes that although his first well-known book with Hammer was widely read, reengineering did not usually lead to success. The reason there was little improvement in bottom line profit in many companies that tried reengineering, total quality management, or many of the other recommended practices is that the tactical changes were not coordinated by an overall strategy. A company's relations with customers, its internal structure, and its relations with suppliers, must be managed in a coordinated way.

A new overall strategy will include new relationships and work processes with customers, with suppliers, and within the company. The strength and the robustness of a company's connections with other businesses, when quality and price of a product or service are not the only parameters of competition, will determine how well it succeeds in today's interactive, fast-changing business world. Cooperation strengthens both the connections with other businesses and the internal operations of the business itself.

EXPANDING INTO THE NEW ENVIRONMENT

Astute business leaders are moving away from downsizing and improving internal efficiencies as the only modes of restructuring. Instead, they are forming interactive links with customers, and bringing suppliers into their own work processes.

> Larry Bossidy, the CEO who is pushing Allied Signal to double digit annual growth, also thinks about the future. He simultaneously prods and pressures his people to become more efficient and to grow into their customers' processes. "Restructuring is negative" he says. "You get a frightened workforce. You need to create jobs."

In industry after industry we see that companies are moving beyond simply providing product and service, and beyond the conception that a customer must be delighted but held at arm's length, to dealing in a coordinated way with the customer connection, with supplier integration, and with internal changes.

R.R. Donnelly, the world's largest commercial printing company, has an interactive relationship with its customers, which has strengthened both its paper printing business and its use of new digital information technology. New, computerized equipment allows the publisher to print promotional inserts or advertisements for specific groups or neighborhoods. In 1995, R.R. Donnelly increased its American and international market share and increased sales by 31 percent. According to E. Wayne Nordberg, an equity partner at Lord, Abbett & Co., a Donnelly shareholder, "They have become [...] masters of their own destiny."

The key to a successful interprise is to coordinate all three areas:
  1) **interacting with customers,**
  2) **integrating with suppliers, and**
  3) **changing internally**

More and more companies are responding to a new competitive environment by proactively linking dynamically and intimately with customers, not only to give those customers the solutions they ask for, but to work beyond that to find opportunities the customer had never imagined existed. In turn, what interprises do for their customers, they expect their suppliers to do for them. They search for reliable suppliers capable of being integrated into their own work processes. In doing so, the internal structure of the company becomes an adaptive, entrepreneurial, learning environment capable of taking advantage of the new opportunities created by the turbulent forces that are shaking up the business environment.

SUMMARY POINTS

⇨ Businesses today are challenged by new competitive drivers that are changing the fundamental way business is done.

⇨ To survive in this competitive environment, businesses are integrating work processes both within and between companies.

⇨ Businesses that successfully restructure to meet the challenge of this new competitive environment are expanding and getting stronger.

THINGS TO THINK ABOUT

☐ Have recent technological advances made a significant impact on your business?

☐ Is the speed of the competitive treadmill in your business accelerating?

☐ Will the rate of change in your business slow down or stop?

CHAPTER 2

THE CUSTOMER CONNECTION

Just as in military history, success in the business world cannot be attained by victories at the tactical level alone. Success requires adoption of an appropriate strategic aim, followed by implementation of suitable operational measures. Interprising companies integrate tactical measures with a strategic aim that takes into account the changing market environment. The strategic aim adopted by businesses seeking long-term survival in this more demanding, competitive, and dynamic marketplace is to become part of each customer's processes. These cooperative efforts have led to a reorganization of the structure and methods of the business.

In addition to selling computer hardware, Amdahl, and many other computer companies also sell information services. They want to work with each customer to understand their issues and contribute to the customer's unique total information process needs. In doing so, they become an invaluable resource, not simply the supplier of a product.

Xerox used to think of itself as a copier manufacturer; now it calls itself "The Document Company," and aims to be part of every customer's printed or electronic document processes.

Remember when bookstores only sold books? The pleasant surroundings and coffee shops in Barnes & Noble and Borders are designed to provide a quiet atmosphere for relaxing and browsing. The aim is to become part of the customers' lifestyles, not just a place for buying books.

CUSTOMER SATISFACTION IS NOT ENOUGH

You hear it all the time, "We evaluate our business based on customer satisfaction rate." "Our customer satisfaction numbers are up five points." It is an accepted and universal truth of modern business, that customer satisfaction is the key to success and security. So, how do you explain the behavior of consumers, 90 percent of whom express satisfaction with their present car, but only 40 percent of whom buy their next car from the same manufacturer? Can you truly afford to rely on customer satisfaction alone for long-term growth?

Connecting with customers solely through sales of products forms a weak bond. When customers are satisfied, but work with you only to buy units of goods, your business can be at risk even though the exterior appears calm. Every day surprised managers and employees may suddenly have to deal with the following dangers:

- The competition reduces margins for their product below profitable values
- Their product is superseded by another
- Their whole technology or business sector becomes obsolete
- A competitor elbows them out by providing a new range of services and added value with the product
- The customer is bought or restructured, and does not need the product any more

Tactical measures, such as sales and price reductions, may increase revenue in the short term, but they do not produce enduring bonds with customers that will ensure long-term strategic success. Increased worldwide and local competition has pushed many companies to reinvent themselves by restructuring and reengineering. What few companies realize is that they must also re-think the way they connect with their customers. The product-only connection no longer guarantees success; **in order to survive today a company must link into its customers' business or lifestyle processes**.

WORLDWIDE CHANGES HAVE DOOMED THE PRODUCT-ONLY CONNECTION

In the nineteenth century, the invention of interchangeable parts by Eli Whitney, together with new corporate structures and distribution systems, enabled a mass production system to arise. At that time, the competitive advantage went to companies that invested in production machinery, and established a hierarchic power structure which ensured efficient use of the machines. In the 1950s and 1960s, the quality and advanced functions of product determined a company's competitive edge. We are all accustomed to this product-based competition which offers customers competitive prices, adds new features to the product line, and lowers costs while monitoring quality.

Increasing worldwide competition has doomed competition based on product alone. Now, a quality product can be made anywhere in the world and high-quality, cost-effective, manufacturing is no longer a competitive differentiator.

There are several reasons for increased worldwide competition:

- Technology and a changing international climate have led to more open competition.
- The standard of education in many countries, including those in the so-called third world, has advanced considerably.
- The cost of production machines has decreased remarkably as manufacturing equipment capability, which used to cost several hundred thousand dollars, is now available for tens of thousands of dollars.
- Sophisticated design aids, usually computer programs, are now available anywhere in the world. For example, professional video production systems have decreased from $200,000 to $20,000 in just five years, while improving their ca-

pability, and are now displaced by inexpensive PC-based computer programs. Computer-aided design and manufacturing software is now available for $1,000 on a personal computer, with a capability exceeding the $5,000 Autocad package of just a few years ago. That package, in turn, did more work more easily than a $50,000 mainframe package sold only ten short years earlier.

- Worldwide information networks make communication and information universally accessible. A company in Sweden can access data on the buying habits and economic statistics for a Brooklyn neighborhood as easily as any marketing company in New York.

It still comes as a surprise to many people that manufacturing capability has become so universally accessible. But consider the recent changes in the printing industry. Two generations ago it was a significant industry, with many people operating printing machines in printing job shops or factories. Today, some projects that previously required a small army of specially-trained operators and technicians are completed by an individual at a desk with a computer, printer, and desktop publishing software. Similarly, in industries from metal machining to baking, high quality production work is accomplished by people throughout the world using affordable new technology.

The worldwide ability to make product led to a political crisis in late 1994, when relations between China and the United States were severely strained. Chinese companies were copying the products of American companies, and infringing their intellectual property rights with significant financial loss to the American economy. They were not only making simple products such as compact disks, but more complex technological products such as Jeeps.

THE NEW STRATEGIC AIM

Modern managers know many operational techniques. They include: total quality management; empowerment of the workforce in teams; reengineering; downsizing; concentration on core competencies; designing better products in less time; reducing the total concept to cash time; and management by walking around. These methods have become widely known as "programs of the month" by employees who have seen ideas come and go, while the company continues to falter. They usually do not achieve lasting success because they are not integrated with a coherent strategic aim.

Before World War II, the French invested enormous resources in the Maginot Line that was to defend their country from invasion. But in May,1940, the German army outflanked the Maginot Line, and conquered France in two weeks using mixed force tank divisions. Their new military strategy was called the blitzkrieg. Like a new military strategy, the new business strategy of moving beyond product and forming process connections with customers must be coordinated with new organizational structures, training, and technology.

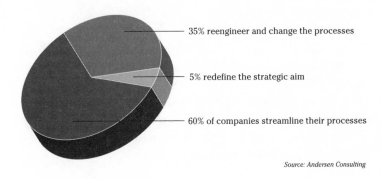

35% reengineer and change the processes

5% redefine the strategic aim

60% of companies streamline their processes

Source: Andersen Consulting

Figure 2.1 How Companies are Responding to Today's More Competitive Environment.

As Figure 2.1 illustrates, most companies streamline or restructure without changing their strategic aim. This is the equivalent of improving the Maginot Line, while the real challenge lies in taking advantage of the offensive possibilities of tank warfare. By becoming part of their customers' processes, businesses can incorporate tactical responses, such as restructuring and reengineering, while establishing the strategic advantage needed to succeed in the long haul.

WHAT IT MEANS TO BECOME PART OF YOUR CUSTOMERS' PROCESSES

Interprises succeed by finding unique ways to become part of their customers' business or lifestyle processes. If the customer is a business, they become part of the customers' business processes. If the customer is a consumer, the interprise finds a way into their lifestyle processes at the end of the value-adding chain.

> Unisys used to think of itself as a computer company; now it sees itself as an information systems company which becomes part of its customers' business processes as it helps them make more money with their customers. They call the process "customerizing."

> Lexus led the way in transforming their business from just selling cars to making Lexus ownership an easy, reassuring experience. The dealers provide coffee and doughnuts in a pleasant lounge while a customer's car is serviced, or pick up the car and return it later, or supply a "loaner." Owning the car becomes a pleasant rather than an aggravating experience.

Business has come full circle. It moved from customer-centered craft methods of the eighteenth century to product-centered mass production methods in the twentieth century. It is now returning to a customer-centered focus. The old craft methods, however, could deal with only a few customers; modern technologies, especially information systems, allow a company to accommodate the unique needs of many customers.

In the past, the invention of mass production led businesses to perceive themselves only as sellers of product, not as facilitators of each customer's business processes. Competition is now leading companies away from the all-customers-get-the-same approach to treating each customer individually, therefore becoming part of the customer's lifestyle or business processes, as interactive interprises.

INTERPRISING HELPS BUSINESSES GROW

Becoming part of customers' business processes can be very profitable.

Johnson Controls was founded in 1883 and has been manufacturing heating systems controls, batteries, and automobile seats, among other traditional products, for over a century. Making product, as opposed to designing it, required mostly "blue collar" work, but little design knowledge. Over the last decade, they have moved from only providing a product such as seats, to providing full design services with the supply of seat systems. In doing so, they have become a tightly connected part of their customers' work processes.

In addition to making controls for buildings, the company also entered the management services market in 1989. They added a full range of services to their products, thus becoming part of their customers' work processes. Group Vice President Terry Weaver described the opportunities in managing the heating, lighting, safety, and security operations of buildings as "explosive ... almost impossible to quantify, a market worth tens of billions of dollars in the United States alone."

CEO James Keyes has said that service now drives the entire corporation and "Most of our growth has come from the fact that we do more for our customers."

HOW TO FOCUS ON YOUR CUSTOMERS' BUSINESS PROCESSES

As we have seen, a standardized approach is giving way to an individualized approach. Advances in communication, computers, and flexible technology are enabling personalized solutions for each customer. Interprising businesses are finding ways to become part of the work processes of business customers, or the lifestyle processes of the consumer customers. Individual attention may come in the form of an individualized product or service, or individualized delivery services, or both. The attention is not limited to the instant of delivery, but is aimed at a long-term, profitable relationship. It is profitable only if the customer feels he or she is constantly getting value from the supplier.

In a desktop publishing firm, such as Kinko's or InstyPrint, the management focus is on service to the customer. In the old printing establishment, the focus was on the workers and the logistics needed to keep the machines busy. The new computer-based printing technology has moved the focus of attention from product to process. Product continues to be the output for which the customer pays, but the focus of management is on the customer's processes. Branch managers do not spend much time worrying about keeping the machines busy, but devote much creative energy to understanding and working with customers. They provide artistic graphic services, accept orders by fax, phone, or computer, deliver their goods, and stay open all night. They are constantly analyzing their customer base and how to extend it. Reliable operation of the computers and printers, backed up by the service of those suppliers, is taken for granted.

There are still advantages for companies that provide competitive arm's length product, marketing, and customer service, but it is only a temporary advantage. Long-term success goes to businesses that not only provide the product and service, but also develop a value-adding, enhancing relationship that binds customers to them. As product and service requirements rapidly evolve, businesses that understand their customers' businesses and help them meet their changing needs will be more valuable than suppliers who supply products and service only.

> During Rome's war against a major economic competitor, Cato the Elder finished every speech in the Roman senate with the words "Carthago delenda est," meaning "Carthage will be destroyed." That was the Roman equivalent of "You cannot avoid your rendezvous with destiny." Company employees and owners should say to themselves, as they walk in to work at the start of every business day, "We *will* find a way into our customers' processes."

SUMMARY POINTS

⇨ The aim of the interprise is to become part of each customer's processes.

⇨ Customer satisfaction with the product is no longer enough to hold customer loyalty.

⇨ Worldwide competition and the rapid proliferation of education and technology have doomed the product-only connection.

⇨ To take advantage of the new environment, companies must restructure their relationships with their customers and suppliers, as well as within their own organizations.

THINGS TO THINK ABOUT

☐ Can a competitor make similar products or provide similar services to those you offer? If not now, could they in two years?

☐ Are you competing on price alone, or do you want to try to contribute to the customer's business or lifestyle processes?

☐ Is your company still focused on products or services, or are you proactively working to understand your customers' processes and to contribute to their goals?

☐ Are your customers asking to have their information systems linked to yours? Do you expect they will do so soon? What opportunities would that open up to provide additional services and become part of your customers' processes?

☐ Do you work with suppliers that understand your needs and help your business do a better job?

CHAPTER 3

THE INTERPRISE PAYOFF

E nabling a customer to make money with their customer cannot be done by sporadic contact at arm's length. To truly become a part of its customer's processes, the interprising
supplier must develop an interactive relationship with the customer. In this new, interactive relationship, each side knows why
they are together: the supplier gets into the customer's processes
to make a profit; the customer needs the supplier's support in order
to make its profit. The relationship is both synergistic and merce-
nary, aimed at mutual business success.

During most of the twentieth century, companies have concentrated on their ability to be world-class in every activity. To be the
best supplier, they organized functionally, eliminated inefficiency,
and sought to have the highest quality product at the lowest cost
and with the quickest response time. It was the customer's job to
figure out how to use the product effectively. Success was measured
by the number of products and services produced and what was
paid for them. That has changed.

> **Changing from being a world-class supplier to becoming a world-class enabler of each customer is the basis for a sustainable strategic advantage in the early twenty-first century.**

The strategy we now see emerging is and must be focused on the customer and what the supplier enables it to do. Since having world-class processes and products is assumed, the challenge now is to go beyond that to enrich the customer. This started as an extension of mass production, and was due to the flexibility afforded by information systems and other technological advances, but it has evolved into a new strategic option for business leaders. World-class quality, cost, and service are simply the entry cards to doing business. They no longer confer competitive advantage. Instead, advantage goes to the organization that understands the customer and its needs, and translates that understanding into measurable value in a relationship that continues over time.

A company does not only supply distinct goods and services, nor isolated bits and bytes of information. Service is not repairing things that have broken or stopped working, but avoiding these things before they even happen. To gain strategic advantage, a company must understand how expertise can be appropriately matched to goods, services, and information over time, and can actually be measured in terms of the customer's bottom-line goals and objectives. Businesses focus, therefore, on mutually beneficial relationships with customers in a variety of classes.

In 1994, when Paul Kefalas became the CEO of Asea Brown Boveri Canada (ABB), regional sales were depressed. Instead of focussing on products or processes, or looking for new customers through traditional product-oriented marketing, Kefalas looked to the leaders in Canadian industry. He asked them to share their corporate strategies and vision of the future. Wherever possible, ABB experts met with representatives of the other company, working with them to come up with ideas that would help them achieve their goals. Less than two years later, ABB has formed over a dozen successful partnering ventures with companies they would never have approached as traditional customers. Sales rebounded, and should continue because of the long-term nature of the partnerships.

THE MASS PRODUCERS

The common wisdom has been that it is not possible to make all decisions in a company from a strategic point of view, so what a business must do is create a set of rules and guidelines and use them to create an effective decision making mechanism. This led many business people to decide that what they needed to achieve was to become world-class. They then concluded that being a world-class supplier meant operating with the least cost, highest quality, and most responsiveness to the customer.

With this goal they could then ask managers throughout the organization to make decisions based on achieving, maintaining, or exceeding world-class standards for their organization. Some went even further to categorize activities carried out in the organization as either being value-adding activities or not. Value-adding activities increased the organizational effectiveness relative to the world-class metrics of cost, quality, and speed of response to the customer.

This has been a reasonable set of metrics, and many organizations used it to focus energies on significant improvements. Several years ago, for example, Hewlett Packard decided that the speed with which new products were brought out was not world-class. Then CEO John Young set a strategic goal to reduce the time it took to introduce new products by at least 50 percent. They succeeded in this challenge, and in the process removed more than thirty committees charged with reviewing new product introductions.

People now ask: What next? How will we compete when we are all world-class suppliers? How will we differentiate ourselves from other suppliers? If all companies supplied the same thing, then they would compete on the basis of the quality of the product—how it works and for how long—the cost of the product, and its availability. For a long while "it" was assumed to be some stable product or service, perhaps even some well-defined combination of products and services. The customers obtained this from banks, insurance companies, car dealers, department stores, shopping centers, supermarkets, even the local grocery store. Commercial and industrial customers obtained their products and services from other companies, retail outlets, or through brokers of various sorts. Even as products have developed a shorter life and a greater variety is offered for sale, companies have continued to choose their suppliers with a fixed vision of a static product.

THE NEW COMPETITORS

Things have begun to change, however. Now companies routinely ask suppliers to treat them more like individuals by modifying the product or service, however slightly or extensively, to suit their needs. This change has been called "mass customization," a blend of the well-understood economies of mass production, with the newer concept that each customer gets his or her own.

> The Custom Foot shoe company has taken a giant step toward mass customization. When a customer enters their store in Westport, Connecticut, he or she will examine the fashionable shoe styles on the wall as a salesperson explains the process, including the store's ironclad guarantee to take back any order. The interested customer will then put on a pair of white socks and thirteen measurements for each foot will be taken by an electronic scanner that transmits the data to the company's factory in Italy. The customer picks out the color and the type of leather, and in two weeks the custom-made shoe will be ready. At around $140, these shoes are priced at the high quality end of the mass production market, but cost far less than the $500 an old-fashioned custom-made shoe would command. In addition to low inventory costs, Custom Foot's flexible manufacturing processes enable it to respond rapidly to new shoe styles.

To succeed, a company must go beyond joining with the customer in producing product, even go beyond linking with the customer's staff and processes to help them with their customers. Going beyond connecting across the no man's land between customer and supplier, the interprise extends and blurs the lines between companies. The supplier teams together with a customer to really understand how their product or service has value to the customer's customer and why. The once closed-doors of a supplier's design operation, for example, must be thrown open to the customer, who joins the process, sharing his intimate knowledge and understanding of the business. We have illustrated this transition in the two drawings below.

Figure 3.1 From Exclusion to Inclusion.

Until the close of the twentieth century, highly competent arm's length suppliers will remain competitive. Those that cannot move beyond that, however, will begin to be replaced not by better suppliers, but by an entirely new type of competitor. These newly emerging competitors have gone beyond being world-class suppliers. We don't know what history will eventually decide to call them, but they are distinct from world-class suppliers in many important ways, and these distinctions will in the beginning provide their competitive advantage. We have called them interprises.

STRATEGIC QUESTIONS FOR STRATEGIC ANSWERS

The most important thing these new competitors know are the answers to the following questions. These deceptively simple questions are not at all trivial; they are at the heart of understanding the change in the competitive landscape and how to gain strategic advantage from it:

a)   What business are you in?

When we ask a collection of successful business people this question, they usually answer in terms of what they supply. If you answered with information about what you do, what you make, or what you sell, you have done the same thing. Instead, extend your concept of your business, particularly toward the way you connect with your customer. In this way, Xerox redefined themselves as The Document Company, and Kinko's now call themselves Your Office at Home.

b)   What do you enable your customer to do?

This is not nearly as easy to answer as it is to ask. Travel agents have answered that they make dreams come true, insurance agents say they sell peace of mind, truck manufacturers assert that they don't make trucks but provide a means of moving heavy equipment over long distances. The Steelcase Furniture Company ran an ad which can be paraphrased "While we manufacture desks, chairs, and tables, that's not what we sell. We sell the ability to share knowledge, to field effective teams, and to discuss ideas."

Figure 3.2 Steelcase As An Enabler of Customers.

The ad shows how they have made the transition from thinking of themselves as supplying what they make to thinking of themselves as enabling the customer to achieve his or her goals and objectives. They have removed their supplier hats and put on the hat of the customer, seeking ways to enrich or enable the customer to meet their own individualized goals.

Answering these first two questions meaningfully is not easy. In leadership seminars for senior executive teams, this question can sometimes generate an hour of tough discussion. Usually team members can easily agree on what they *supply*, but understanding and describing what they *enable* is a new and more difficult concept. Experience has taught us that although a company might supply a cohesive and easily described collection of products and services, there will be customers who use these products and services in significantly different ways.

Companies are not accustomed to thinking about the ways in which their products and services bring value to the customer. When they articulate these thoughts, it is usually necessary to follow up with further questions, such as: How do you know? or How does your customer measure the value you bring? These questions are not designed so much to be answered as to shift the pattern of thinking.

Mass production taught everyone that to be world-class was to be the best supplier of goods and services, inferring that all customers have the same needs. **The assumption that all customers have the same needs is now false**. This means that before understanding what you enable your customer to do, you should first ask:

c)   How many different kinds of customers do you have?

As you think about answering this question, ask yourself the following: How are you going to identify classes? By what discriminating factor? Why is that factor the one to use? What choices do you have?

Early experience shows that most executive teams can sort customer classes using the principles of market segmentation. They identify different usage of the products and services by market segments, and consider each market segment to be a class of customer.

Airlines, for example, readily identify business travelers and vacationers as two broad segments. They then begin a process of further segmentation and come up with subcategories.

This approach does not go far enough in seeing customers as individuals, and it persists in herding them together in averaged groups, smaller than before, but still groups. It does not get to the nub of the issue, which is to understand your customer's context and inner world. To get away from this thinking, try to answer questions such as: What brings your customers back to you for repeat business? If your customer had a wish, what would he or she want you to change? These type of questions help classify customers by their degree of sophistication and familiarity in doing business with you or your industry. It is also useful to have others critique your set of customer classes, since they do not have an insider's tunnel vision.

Many manufacturing companies solve technical questions using the Japanese principle of Kaizen, which requires that you ask the question "why?" five times, penetrating deeper each time, like peeling the layers of an onion, until you uncover the core issue. The same principle can be usefully applied to understanding the constraints and motivation of your customer. Why do they want a particular product or service in the first place? What do they do with it to make their profit? What else is added to it? How is it marketed? Why? Why do they succeed?

d)   How does your customer measure the value they receive from your relationship?

Before answering this, think about how you would like the customer to measure the value they receive from their relationship with you. Then ask who is best able to say what the customer will value. The answer to this last question is obvious—the customer.

An aviation fuel company once told us the most valuable thing they can provide to an airline is pure and clean jet fuel. They talked at length about the problems impurities would cause an airline. A customer of theirs who heard this responded by saying that they assumed this was the case for all viable suppliers; they would not consider buying aviation fuel from anyone whose fuel was not pure and clean. They chose suppliers who could most reliably ensure that the airline could land a plane and refuel it within fifteen minutes. The airline had calculated the value of this relationship in terms of the substantial costs for every minute of delay in the refueling process, while the supplier made its own—false—assumptions about the customer's value criterion.

e)   Do you ask your customers to help you help them?

Most companies will send people to ask their customers what they want. They meet with the customer and report back on what they think the customer had said. This is not good enough anymore. Your people must live with the customer, in the real or cyber world, learning their business from their perspective, figuring out how best to convert your skill, expertise, and information into goods and services that will have a measurable impact on their bottom line. In this way, the customer feels value from the relationship itself, which allows you to anticipate opportunities for mutual advantage that even the customer may not have foreseen. This is the interprise payoff.

SUMMARY POINTS

⇨   The direction of the changes we are experiencing is:
>   From arm's length transactions to the customer connecting with supplier and working together to create goods and services.
>   From a focus on commodities to a focus on specialty items combining physical goods, information, and service in a relationship over time.

⇨   There are a series of questions which can help you focus on the change from being a world-class supplier to an enabler of customer enrichment.
>   What business are you in?
>   How many different classes of customers do you have?
>   What do you enable your customers to do?

⇨ A secret to success in 21st century competition will be to succeed in working with customers to help you help them.

———————————————————

THINGS TO THINK ABOUT

☐ How do you interact with your customer? Do you let them into
your processes, such as planning, design, distribution?

☐ What value could you bring to your customer that would make
it worth their while to let you become part of their processes?

☐ What do you supply that your customer values most?

CHAPTER 4

DYNAMIC RELATIONSHIPS

D riven to a new level of competition, companies are reevaluating production and service cost *down the line*. The fraction of the total cost of service or manufacturing process internal to a company is usually between 5 and 20 percent of the total cost. The other 80 to 95 percent is due to the other companies in the value-adding chain. Internal efficiencies, therefore, only go so far.

To achieve true cost efficiency today, companies need to define and manage processes across the entire value-adding chain. This can be done only when companies coordinate the work processes between them. Pressuring a supplier to reduce prices, while keeping him at arms length from your work processes, cannot match the improvement that is achievable by coordinating those processes.

Wal-Mart expounds a win-win-win business philosophy (customer wins, Wal-Mart wins, supplier wins). The retailer willingly allowed the manufacturer of Wrangler jeans to take the responsibility of controlling the supply and restocking of its jeans to all Wal-Mart stores. This required monitoring sales on a store by store basis, and watching which sizes and styles were moving at each store. The result was a threefold increase in the quantity of Wrangler jeans sold by Wal-Mart. The customer was offered a wider variety of jeans, and both Wal-Mart's and its supplier sold more goods.

THE ARM'S-LENGTH, SLOW, OLD WORLD

Modern research of businesses as systems shows us that:

- In the past, business processes were isolated between companies; now they are coupled.
- In the past, business processes were static; today they are dynamic. Management decisions are affected by the velocity, acceleration, and inertia of business processes, both in their own and in connected businesses.

The figure below shows businesses passing goods one to another, as they have for the last century, when mass production and mass distribution systems reigned supreme. In that world, a customer took no interest in how a supplier did its work, so long as the product was in the warehouse or the service available when needed. There was no need to treat each customer individually, nor could you afford to do so even if you wanted.

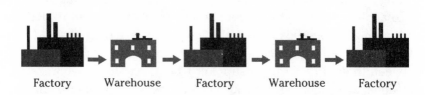

Factory   Warehouse   Factory   Warehouse   Factory

Figure 4.1 Traditional Handling of Goods.

Change in this system was a slow process. Prior to altering the flow of goods in the value-adding chain, the goods in process and in the warehouses needed to be used up, or else written off. In a world of isolated intercompany processes, you could successfully manage a business by carrying out a single change at a time, and each improvement effort would move a company ahead for several years.

THE INTERACTIVE, DYNAMIC, NEW WORLD

The interprise, however, introduces change as a coordinated set of activities. These include what it does for customers and how it interacts with them, how it organizes its relationships with customers to support their business or lifestyle processes, its internal organization, and its interactions with suppliers.

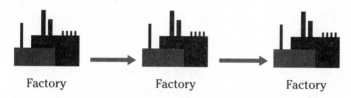

Factory   Factory   Factory

Figure 4.2 New, Coordinated, Work Process.

Figure 4.2 shows this evolving new world in which interprises link their work processes. This system is not only coupled, but is capable of rapid change, since the amount of inventory in the system is small enough to permit dynamic behavior. This is how Wal-Mart and many other modern stores work with their suppliers.

THE DIFFERENT BEHAVIOR OF LINKED, DYNAMIC BUSINESSES

The behavior and management of a dynamic system, in which factors such as requirements for products, or the rise and fall of market opportunities which accelerate and decelerate, is fundamentally different from that of a slowly changing system, where the rate of change is barely felt. Not only do influences ripple through the whole system from one company to another, but the timing of those ripples affects decisions. You cannot manage a linked dynamic interprise system with the rules and methods we used to manage businesses in a slowly-changing, decoupled business world.

The air traffic system in North America is a dynamic linked system. The passengers are the equivalent of inventory processed by airplanes. When bad weather or computer failure slows or stops landings at Chicago's O'Hare airport, the whole system is thrown into disarray. Flights all over the U.S. are delayed or canceled. One flight affects another, and the timing is critical. Airlines schedule planes to minimize the time passengers wait, rather than maximize the occupancy of seats. The other method, filling up the planes but keeping the passengers waiting, would eliminate dependence between flights. Schedules would have long waits between flights, so that even if flights backed up, the airplane would never be kept waiting, but the passengers would. That is obviously not the way to run airlines. It is, however, the way many businesses are still managed.

Many businesses have not yet evolved with the customer to support the customer's processes. They prefer to keep the machines or workers busy, and the delivery trucks full. For them, capacity utilization is the guiding principle. This requires sufficient levels of work-in-process and inventory (analogous to waiting passengers) to keep the resources, both people and machines (analogous to crews and airplanes), busy. While many companies are still managed by those old rules, the business environment has moved from static producer-pushed product orientation to dynamic customer-pulled process orientation.

THE BUSINESS WORLD HAS BECOME DYNAMIC AND LINKED

The introduction of new products has accelerated enormously in the last ten years. The figure below shows the product technology change frequency for automotive steering systems at General Motor's Delphi Saginaw Steering Systems. These rates are typical for many industries, forcing companies to react quickly and dynamically to innovative technologies and a demanding marketplace.

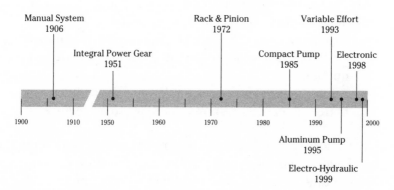

*Figure 4.3 The Accelerating Rate of New Product Technologies.
Automotive Steering Systems as an Example.*

Both in individual businesses and in the business environment, examples of new dynamic and linked systems abound.

> Motorola, faced with erratic swings in demand for its products, has developed a method of sharing schedule forecasts with its suppliers. Suppliers are asked to keep inventory within stated maximum and minimum levels, and Motorola connects its information systems, with the anticipated demands of supplied product, to those of the supplier. Suppliers then have the freedom to plan their work to take account of all their customers while maintaining product levels within Motorola's limits. The supplier, rather than Motorola, is responsible for the analysis leading to order quantities. This approach has significantly lowered costs for several reasons:
>
> • Motorola's number of purchasing positions has been cut by about 30 percent since inventory analysts are no longer needed
> • Suppliers need fewer people to analyze Motorola's plans

- The number of inventory turns has doubled, indicating an increased efficiency of the system
- Suppliers who supply a mix of similar parts reduced costs because of more careful planning of their production
- A supplier who serves several Motorola facilities can better allocate its resources across those facilities

In dynamic business systems, the time delay between events has a critical effect. The width and positioning of windows of opportunity, or the time delay between receiving an order and fulfilling it, are critical to success. While businesses have moved from being isolated and static to being linked and dynamic, our management systems and methods have not. So here we are, trying to manage linked dynamic systems, while using methods, traditions, and mind-sets of a bygone, decoupled, static world. No wonder life is traumatic!

The infamous Wall Street collapse in October 1987 was due in part to the imposition of linked dynamic effects on a system designed to deal only with decoupled static effects. This was caused by the technology of program trading. Brokers' computers, linked to the stock exchange computers, tracked prices and trends, then bought and sold. Many brokerage computers, using similar rules for buying and selling, noticed trends simultaneously, then all sold together, exacerbating the fall.

The solution to the problem was to install a system of 'circuit breakers.' For instance, if there is a fifty point change in the Dow Jones index, arbitrage trading is suspended. These measures were designed to disconnect the linked dynamic computer system in the event that a rapid acceleration of trends is detected, thus forcing the system to behave as a decoupled, unaccelerating, static system.

Such dynamic, coupled behavior is, however, becoming the norm in many businesses where there is no regulating body which can order intercompany transactions to decouple and slow down. We need to deal with the new environment and use it to gain a competitive advantage. And we can. Large companies like Motorola, General Electric, Allied Signal, and Coca Cola, and smaller ones like Solectron, Iscar, and others mentioned in the book, are becoming interprises, succeeding in a business world characterized by dynamic interaction. The leaders groped their way to success. By describing the new business world and the variety of measures taken by diverse companies, any business can learn from the experience of those leaders.

SUMMARY POINTS

⇨ When moving from arm's length to interactive relationships, work processes speed up and become dynamic.

⇨ Dynamic systems behave in ways that are fundamentally different than static systems.

⇨ Management methods are based on static behavior. As the business world becomes more and more dynamic and interlinked, static methods no longer apply to the new reality.

THINGS TO THINK ABOUT

☐ Does the window of opportunity come more rapidly than you can accelerate into it, then go away more rapidly than you can decelerate your business?

☐ Are you turning away customer requests because you cannot react fast enough? If so, where do those customers go? Do you have a competitor who can accelerate and decelerate better than you?

☐ What measures could you take to give your company less inertia?

CHAPTER 5

ADDING VALUE BY SUBTRACTING TIME

For an interprising company, reducing production time is more than just a technique for saving money and getting product to the customer faster. It offers the possibility of restructuring the organization to discover new market opportunities by penetrating customers' business or lifestyle processes, while increasing integration with suppliers.

Time is being used as a competitive weapon by the U.S. textile and other industries to enhance their positions in international trade.

The industry development group Textile and Clothing Technology Corporation [TC]^2 in Cary, North Carolina, is working with companies and research organizations to improve the competitiveness of the American clothing industry by using time as a weapon. In 1992, [TC]^2 set a goal of reducing, from sixty-six weeks to three weeks, the total processing time for all the value-adding chain–from the weaving of cloth to the sale of retail product. By 1995, its Quick Response project had been integrated into the chain of companies and stores. They had reduced total process time to thirty-six weeks, and were well on their way to the three week goal.

To achieve a three week response time, businesses along the entire textile "food chain" must interprise. They need to work closely with the customers' people, sharing technology and knowledge in order to rapidly produce the finished goods for mutual benefit.

TIME COMPRESSION OVERWHELMS THE COMPETITION

Time compression is leading to an entirely new environment in the textile and garment industry, as it moves from "make to forecast" to "make to order" methods. A make-to-order system will rapidly provide a customer with a custom-tailored garment. Until now, hardworking, low-wage tailors in the Far East have made clothing to order, but they offer a limited range of styles and cloths. Soon we will start to see widespread customization of clothes with a great range of variety, quickly produced by well-paid workers using automated machinery. Levi Strauss already provides custom jeans, charging an extra $10 per pair for the customization. The high value clothes will be custom-made, as traditional mass production is pushed to the less profitable end of the market.

The textile industry is just one example of how dramatic reduction of response time goes beyond changing the playing field; it creates a new game with new rules. Competitors will face two alternatives, both problematic:

- Keep large inventories of the many styles and sizes needed to satisfy the fragmented market (see Chapter 8, "Divide to Conquer"). This adds considerable cost. In the garment business, the cost of inventory is 27 percent of the retail cost of the garment. Comparable numbers for inventory cost exist in different industry sectors; or,
- Make clothes as rapidly as needed, pulled by orders, not manufactured to forecast. For overseas competitors, their entire textile and garment chain would have to metamorphose into a cooperating, interprising unit. They would still face high air shipment costs.

By creating competitive advantage through time compression, you force your competitors to make difficult choices.

VALUE-ADDING PROCESSES

Value focus sets the price of a product or service according to what the competitive market leads the customer to pay for it. Cost focus looks at products or services from the perspective of what they cost to provide. The value of a product or a service is the value in the eyes of the customer, reflecting the amount that he or she will agree to pay for it.

Work processes are considered as value-adding and non-value-adding. Value-adding time is time spent doing things for which the customer will pay. For instance, the customer will pay for any processes involved in making the product, such as cutting metal, boiling chemicals, or sewing. Non-value-adding processes include waiting time, transportation time, distribution time, or time spent testing the product. These are items which add no value from the customer's perspective.

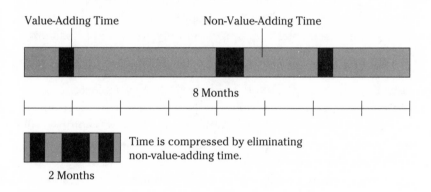

Figure 5.1 Value-Adding vs Non-Value-Adding Time.

In many industries, non-value-adding time may be as much as 95 percent of the time from when the customer orders the goods until they are delivered and accepted. It follows that time compression is achieved in the first place by reducing or eliminating non-value-adding processes, and only then by investing in speedier value-adding processes.

Quick response manufacturing in any industry reduces costs down the value-adding chain and creates an insurmountable competitive barrier to a "make to forecast" company. In addition, it often overcomes the advantage of cheap labor rates. As we read every day in the newspapers, the environment of "make to forecast" plays to the strength of a low-wage, educated labor force found in many regions of the world. Quick service or production is a key to modern competitiveness. Whether or not your own company is moving in this direction, your competition already is.

THE RACE IS ON

When one company accelerates, others must try to keep up. Development of a new automobile model once required five or six years, with only styling and subsystem changes during that period. Toyota led the way in developing new models in two or three years, and now every automobile manufacturer is cutting down development time. Chrysler's award-winning Concorde, Vision, and Intrepid sedans were developed in approximately forty months. Ford is working hard to reduce the development time from four to three years. When Toyota set the pace, the automobile industries worldwide sped up, even as Toyota made their processes faster.

Interprising companies that integrate their business processes with those of customers and suppliers can respond rapidly to personalized production requests. As the marketplace assimilates and expects more personalized products or delivery processes, traditional mass production will become less competitive. Personalized products cannot be made to forecast. When customers need customized products or service quickly, they will turn to the fast, interactive, interprising company that understands them and is already a part of their business.

INCREASING PRODUCT VARIETY

Although a traditional company tries to understand the needs of its customers, its operating processes are not designed to change whenever the market or customer needs change. Its information system is usually not capable of following the individual desires of each customer and potential customer. Employees and organizational structures do not welcome sudden changes in market conditions as an opportunity for finding new markets and methods. In an interprise, all interactions with customers are tracked individually and rapid change is dealt with as a matter of routine. The company can rapidly introduce a great variety of products.

An interprise benefits from changes in the market and introduces new dimensions to competition. An interprising business uses time response, product differentiation, geography, or distribution channels and methods to track a fragmenting market. Interprise means developing an adaptive, organizational structure that continuously finds new, creative solutions. Even better, an interprising business develops new products and services and leads the move to fragmentation. It takes market share from its tardy competitors, as Nike did for sports shoes, Swatch for watches, and RayBan for sunglasses.

RAPID PROCESS FOCUS CHANGES COMPANY STRUCTURE

If the idea of speed is so good, why isn't everyone doing it? Speeding up business processes requires the restructuring of a company, whether by empowering cross-functional teams, encouraging the entrepreneurship of individuals, or leading by coaching rather than bossing. Time compression efforts fail when companies try to speed up processes for a particular project but neglect to set a goal of becoming an interprising organization. Management must permit and encourage these changes, rather than focusing on product while neglecting process reorganization.

The successful restructuring at Ross Operating Valves of Livonia, Georgia, shows how time compression efforts led to empowerment of individuals and profitable integration of Ross workers into customers' development and operations processes.

In 1985, Henry Duignan was appointed chief operating officer of Ross Operating Valves, a long-standing manufacturer of pneumatic valves. Russ Cameron, the chairman and sole owner, had turned to Duignan when the company was feeling the strains of decreasing margins within a difficult competitive environment. As soon as the company was on sure footing, Duignan, with Cameron's support, built a quick-response, "make to order" capability which Cameron called RossFlex. Prior to this, Ross had sold only catalogue products. A customized order had required a year to produce and cost over $100,000. They now can produce a customized valve within five days for $3000.

How did they do this? Duignan invested in some computer-aided manufacturing equipment and design systems, but the most significant change was in the organization of the people. Like most companies, engineers and technical personnel had previously been in functional departments, such as designing, manufacturing, and purchasing. Duignan hired young engineers and told them they each would personally deal with every aspect of a customer relationship, from specification through field installation. No complex workflow organization or complex factory reorganization was provided.

The computer-based design and production capability empowered the young engineers who sat at the controls of the computerized "RossFlex" system and they quickly found ways to interact with the customers' engineers and solve the customers' problems. They achieved results that previously would have required expert designers supported by the infrastructure of a design department, and expert machinists supported by a production department.

This was a major internal change. As a result of quick-response, customized solutions, customers moved from seeing Ross as providers of catalogue valves—and there is always more than one company happy to provide catalogues and products—to seeing Ross employees as part of their problem-solving team. The customer's engineer, who once waited a year for the customized valve, worrying every day that something was not correctly and completely specified, now gets the valve within a week. He

tries it, asks for modifications, and after a few short interactions, they come up with a solution in which the customer also feels ownership.

This process usually does not even require the formality of
a purchase order. The customer calls "his" engineer at
Ross, they work on a problem, and the bill gets paid. The
relationship is like that of a service provider. You do not
take a purchase order to a doctor or a lawyer, for example,
nor in this case to the team member, to solve your technical problems. Manufacturing thus becomes a service.

Ross has benefited in two ways. Word-of-mouth has brought
them so much work that customers are lining up. Ross are expanding as quickly as they can. Furthermore, according to Henry Duignan, they have "mined" their customers' minds, learning so much
from them that new, profitable product lines have been created for
their catalogue valve factory.

Ross sought innovative answers to decreasing margins and increasing competition. By seeking overall solutions rather than focusing only on speed, the goals and outlook for the company
drastically changed. Ross became an interprising organization.

FROM ISOLATION TO TIME COMPRESSION

The much-reported Ross Operating Valve story illustrates the dramatic ways that time compression changes both a company's internal operating processes and its relations with its customers.

- The company's goal moved from selling catalogue products to
  entering the customer's problem-solving processes.
- Time compression, which led to the customized solutions, required the empowerment of knowledgable and motivated people.

Currently, most businesses are still product-focused. The design
department is given missions, product by product. The goals of
manufacturing are defined by product: produce 100 model A wid-

gets and 500 model B widgets next week. The mission of salespeople is defined by product, and sales targets are listed for each product or product family.

A product focus leads to isolation not only of the processes between companies, but also to isolation of functions within companies. The product designers and makers have little or no influence on the marketing and production forecasts. They work according to the sales department's agenda, producing goods for a warehouse, with all the lack of motivation implied. Workers do not care who the specific customer for the product may be because it has no impact on their work. They do not care who supplies the materials and components because the purchasing department handles orders, forcing them into isolation from their suppliers.

Time compression is a powerful technique for companies that want to move from isolation to cooperation. Employees can be motivated by the challenge of reducing time because the competitive goal is easily understood. Korean shipbuilders build a general cargo ship in fourteen months, American companies need twenty-eight months. The time challenge to the U.S. shipbuilder is clear.

Time compression efforts produce increases in communication, knowledge-sharing, and creative problem-solving. They advance other important issues, including redefined aims for a business, reorganized partnerships with suppliers, and new structures and methods within the company. You can start by limiting the aim of time compression to just that—reducing time. The goal then changes to that of increased integration with the business processes of customers and more cooperation within and without the organization, becoming an interprise.

HOW SHORT IS YOUR "HALVING-TIME"?

"Halving-time" is the time required to cut in half the time taken for work processes. After one halving-time, work which previously required a year would take six months; after two halving-times, the work would be completed in three months.

An amazing number of companies are achieving short halving-times of eighteen months to three years, and in doing so can expand into new markets which provide stimulating challenges to their expanding creative workforces.

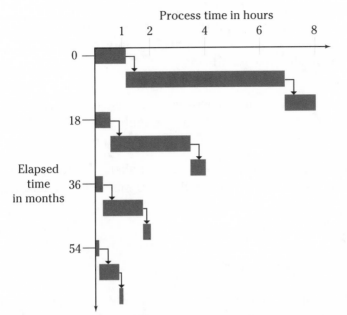

After 3 halving periods of 18 months each, the time
required for these processes is 1/8 of the original time.

Figure 5.2 Halving-Time.

Rick Pyfrom, plant manager of Johnson and Johnson's
Vascular Access Division, which holds just over 50 percent of the market share for catheters, has been running
a successful improvement process in his plant. The goal
is to reach a lead time for production which is not more
than ten times the value-adding time (while doubling
quality achievement metrics), at a rate of 2 percent per
month. This will halve the time needed for manufacturing
his products in less than three years.

Years of hard work and investment have made Motorola a
company with an enviable reputation for quality and for
moving with the times. The reject rate in product production is now down to a few parts per million. Where
does the company go next? Since the early 1990s, their

people have faced the 10X challenge; to reduce the time taken for every work process by a factor of ten in five years. This is equivalent to a halving-time for every work action of eighteen months!

Ingersoll Rand, manufacturers of heavy mechanical equipment, is successfully implementing a challenge for their people to halve times every eighteen months.

Interprising companies take the time challenge seriously, both as an end in itself, and to drive the structural and management changes they know are needed for survival in today's intensely competitive world.

ORGANIZATIONAL INERTIA

Automobile advertising used to state proudly the top speed a car could reach. These days they advertise the acceleration—0 to 60 mph in 8 seconds. Now that everyone can go fast, the differentiator becomes how quickly you can hit top speed. A large truck may be able to go faster than a small car, but the small car can accelerate and decelerate more quickly then the truck. This is, of course, because the car has less inertia. As the pace of work speeds up, the old phrase, "inertia of the organization," takes on a new and critical meaning. Companies with more inertia, where change takes time to penetrate, lose out to more nimble companies with less inertia. Therefore, time compression is necessarily related to the structure and inertia of the organization.

In an interprise, time compression works together with the other key interprise factors we are discussing; namely, penetration of customers' business or life style processes, leading to larger and novel markets; empowerment of workers; and new relationships with suppliers.

Quick response to customers, changing markets, and new opportunities cannot be achieved when functions are divided into isolated departments or when most decisions require management approval. Quick response companies inevitably develop an adaptive corporate structure with self-managed work teams. The new structures influence investment in technology (especially communications technology) change performance measurement and pay structures, and change financial systems and cost accounting procedures, that often cause wrong prioritization and harmful decisions.

This changed internal company structure is the subject matter of the third part of this book. An interprising, quick-response company moves to a different competitive game. It is structured around interactive processes within the company and between companies. Competition becomes multidimensional, opening up new and novel competitive opportunities into which it expands.

SUMMARY POINTS

⇨ The time needed by companies to halve time for each work process varies from eighteen months to five years.

⇨ Time compression provides a powerful competitive weapon.

⇨ You reduce time primarily by eliminating, or reducing, non-value-adding time.

⇨ Time compression restructures relationships with customers and suppliers, and within the organization.

THINGS TO THINK ABOUT

☐ Does your company have a challenge to dramatically reduce time to do work? What is your halving-time?

☐ For which services or products will a dramatic time reduction increase your attractiveness to customers? How much time reduction is needed? Half, or more?

☐   How much time reduction do you need to achieve to move ahead of your competitors?

☐   Are your competitors thinking about compressing time in order to gain market share?

CHAPTER 6

PRODUCTS AS PLATFORMS

T he *Fortune* issue dated May 15, 1995, which listed the Fortune 500 companies for 1994, was an historic watermark. Previously, the publication had been listing the 500 industrial companies separately from the 500 service companies. For the first time, companies of all kinds—manufacturing, service, distribution—were combined in one list. *Fortune* had bowed to the inevitable: industrial companies were no longer differentiated from businesses selling service or knowledge.

The pervasive spread of technology and the availability of quality products which has led to increasing expectations of customers, have created many more product producers. Competition based solely on hardware product price and function has become more and more cutthroat. In order to beat the competition, producers now add information and service to their product. Today, electronics and information are both components of a product and the bedrock of service systems. This leads to reconceptualization of the word "product" to mean a fusion of hardware, software, and service, giving a total customer solution. The hardware "product" is no longer an entity standing alone. It has become a platform for maintenance, information, and service, providing value to the customer's processes over a long time frame.

Mack Trucks, Inc. provides examples of several aspects of interprise capability, and these appear throughout the book.

Chief Operating Officer Donn Viola is justifiably proud of the way Mack Trucks is staying abreast of the computer revolution. Electronic information systems have changed not only the manufacturing systems which make the truck, but also the truck itself. Electronically controlled engines and computer diagnostics allow adjustment of the engine and suspension to match changing conditions of freight, altitude, and terrain, bringing significant savings in operating costs. More than thirty vehicle management functions can be programmed in the system, including fuel control, torque limiting, idle adjustment, road speed limiting, and horsepower rating.

A display above the instrument panel gives the driver a real-time report on the vehicle's performance, and if there are problems he or she can radio ahead to the service facility to minimize downtime. The truck arrives at the service bay with the problems diagnosed, and the tools and parts already waiting. The truck, in addition to being a physical product, has become a platform for electronically based maintenance, information and upgrade services.

Courtesy of Mack Trucks, Inc.

Figure 6.1 A Mack Truck Service Bay.

SUPPLYING HARDWARE ALONE IS NOT ENOUGH

What benefits do the manufacture and sale of a hardware product bring to the supplier and the buyer? Product used to mean hardware sold to a customer for a one-time sum of money in a single transaction. Now a supplier sells not only the hardware, but a fusion of hardware, information, and service, in which the hardware is the platform, a basic component of the mix, but no more than a component.

A piece of hardware by itself, without information and maintenance, is never a solution to anyone's problem. In the past, when an industrial customer bought a component for a product it manufactured, it performed the engineering and other value-adding services necessary to incorporate the component into that product. In the case of consumer products, stories abound of automobile or household appliance repair businesses which, intentionally or not, perform bad work for too much money. In the modern era of the interprise, the supplier takes responsibility for the servicing and maintenance. Product now comes bundled with all the information and service needed to facilitate its maintenance and use, and to make these more reliable. If your supplier does not provide it that way, you will soon find one who does.

A decade ago, the automobile industry, as in many other industrial sectors, moved the design work to the suppliers, taking advantage of the supplier's expertise for the components it made. The requirement now is for suppliers to work together to produce a complete subassembly. The big car company customer is no longer satisfied with product alone. It wants product to include all the information, design, and integration services.

HOW INTERPRISES ARE FUSING
PRODUCT+SERVICE+INFORMATION

Product design previously focused only on the function and price of hardware. The spread of networked information as the basis for ongoing maintenance and other services has expanded the scope of product design to include these activities. Design work now incorporates into the product the ability to support maintenance and other services over the life of the product. The services include de-commissioning and recycling of the product at the end of its useful life. Just as *Fortune* magazine cannot distinguish anymore between industrial and service companies, so the boundaries between man-ufacturing and service have disintegrated.

Goodyear has gone beyond supplying a tire as a piece of product hardware, to supplying the tire as a service. The tire has become a platform for a long-term, mutually sustaining relationship between the customer and the interprise. Digital information infrastructure is a key to their logistics system, which supports supplying tires as a service.

> In March 1996, Goodyear launched the long lasting Infini-tred tire with a lifetime offer: the car owner pays nearly double the cost of a regular tire, for which Goodyear guar-antees to replace a worn set of tires free within 3 years, and at half price thereafter. Some industry watchers were surprised by this move, but it followed logically from Goodyear's experience in the airplane tire business. For years, Goodyear has been taking care of every aspect of supplying and maintaining airplane tries, charging a fee for each landing.

> The digital information infrastructure is a key to Goodyear's logistics system which supports their supply-ing tires as a service, not as a stand alone product.

The fusion of product, service and information penetrates every field. This next example shows how the product, medicine, is used as a platform for value-adding information. It also shows how com-panies interact to do so.

> Bergen Advertising are working with Apple Computer to give pharmacists a computer with a CD ROM or online hook-up that provides complete information about medi-cines, generics, and costs, so that the pharmacist can choose the best solution for a patient. While the pharma-cist searches for data, the system also displays attractive advertising relevant to the medicine type being searched for.

AT&T advertises a service to help small businesses with start-up problems. They have changed focus from supplying communication capability alone to generalized business consulting, because that meshes with their customers' processes and builds on their capability in communications systems.

Order fulfillment systems in companies used to deal with each order separately, tracking limited information about delivery schedules. A modern order fulfillment system deals with the total customer history, and anyone using the system can see every interaction anyone in the company has had with the customer. There is no need for the customer to be shuffled from one person to another, explaining at each step what the previous interactions had been. This follows the lead of service organizations. For example, travel agents include in their databases all the customer's preferences for air travel, hotels, car rental, and all the other items which are part of the travel process. Car rental companies, hotels, and airlines remember in their computers the personal preferences of each person registered. A decade ago your personal travel agent may have remembered your preferences. Today they are held in a computer so that anyone dealing with a request, at any network-connected branch of the company anywhere in the world, has all the information presented whenever a transaction is processed.

More and more products are appearing with the benefits of information technology built into the hardware. Industrial machines, such as metal cutting machine tools, large computers, or printing presses, often include sensors. They monitor the performance of the machine, and communicate automatically with the service organization, diagnosing malfunction and organizing service, so that the user or owner of the machine need not pay attention to maintenance. Often the repair can be achieved by electronic communication, without the user even knowing there had been a problem. Scitex and Otis, two well-known corporations, are examples of companies which have taken system maintenance to new heights.

Scitex is a $600 million a year company, a leader in color image communication systems serving the publishing, broadcasting, and paper and textiles industries worldwide. Seventy percent of its development and production are carried out in Israel; there are distribution centers in Boston, Brussels, Tokyo, and Hong Kong. Users of Scitex systems routinely share files among personnel linked by computer networks, and each system is configured to the work requirements of individual customers.

Scitex operates its own global computer network that enables interaction between customers and Scitex sales personnel, between product developers and production schedulers, between management and all other units of the company. The customer can access documentation for its system from any location through the Scitex network. Maintenance is facilitated by direct communication between the customer's system and a maintenance center. Customer application experts maintain a database of customer problems and the impact of proposed solutions, which is shared through the network. The product has become a platform for service and information.

Otis Elevator has installed OTISLINE, a centralized dispatching center linked to an integrated corporate database that includes customer data and preferences, customer service histories, lists of all service mechanics and field offices, as well as constantly updated information from actual field service experience with Otis products. Its twenty-four-hour customer hotline handles half a million calls a year, and tracks all calls to completion. Management can monitor product performance data in real time and field offices can follow all service activity in their region or in other regions to see emerging patterns of problems.

A major selling point of this system is the ability to show potential customers "raw" product performance and service response data. The system has also eliminated paper service reports; instead, service mechanics use hand-held computers linked by radio to a private network. The network distributes routine service data to engineering departments for product performance improvement, and ideas for new product capabilities. It is also connected to remote elevator monitoring sensors and diagnostic technologies for proactive maintenance and response to problems before they are reported. Otis has turned their product into a platform for service and information.

USING THE PRODUCT PLATFORM TO EXPAND BUSINESS

Packaging provides an opportunity to add information to a product, adding value by specializing or localizing it, for example, having the store's name on the packaging of supermarket products, or local advertising on Coca Cola cans. When a store establishes a bulletin board for local citizens, it is using the free distribution of information as an attraction to customers. Ingenuity in any business leads to successful ideas for using information to make product plus information plus service combinations more profitable.

The Mayekawa organization in Japan makes and supplies industrial systems, mostly cooling systems. Their engineers do not satisfy themselves by merely getting a specification, then supplying a system; they insist on understanding the customer's entire problem, often reaching a better understanding of that than do the customer's employees. Among their products are cooling systems for the food industry. They understand that food is delicate, that fish needs to be treated differently than vegetables, and that taste and nutrition are affected by all the processes the food goes through, from harvest to the food table. Because of this, Mayekawa's people develop each system individually, making sure they understand the key problems in the entire processing chain. Doing so has not only led them to long-term interactions with each customer, but to entire new businesses.

For instance, Mayekawa worked with a cooperative association in Abashiri and its distributors and suppliers, to improve the freezing process for scallops. From this interaction came a machine to shell and slice scallops automatically, then a machine to sort scallops automatically for the sashimi, cannery, and consumer store markets. Mayakawa's people started to talk about scallop processing with fishermen, distributors, cooks, and others in the scallop industry. Their people lived and worked with scallop distributors and processors. They concentrated on improving the total scallop process, from the time scallops are caught until they are eaten.

Mayekawa is an expanding organization. They use product as a platform to interact with customers, to develop long-term relationships, and to uncover new opportunities for business expansion.

Changing the hardware of a product is expensive and usually not possible. Software, on the other hand, can be modularized and changed easily by reading a new copy of the software into the equipment. The clever use of software in a product can help retain or expand markets.

A company which makes and installs engineering hardware systems that include software, needed to protect their business in China and other newly industrializing countries. (They wish to remain anonymous.) Assuming that their customers may copy their product then send them away, they took two defensive actions. First, they designed the systems so that maintenance would need special software. That software is not included in the product as delivered, but is a nonpurchasable component of the instruments which they use to service the product. Second, they planned a series of software upgrades so that when the customer copies the product—which they regard as inevitable—the copy will be unmaintainable, and unupgradable, and therefore uncompetitive.

ADVANCED TECHNOLOGY AND FUSION PRODUCTS

We have seen how, with the aid of modern technology, service and information can be built on the product platform. With business acumen, this can be transformed into a significant part of the total customer solution, whether the business sector is toys or airplanes, simple products or expensive ones. On older high performance airplanes for example, the pilot would pull and push levers to control the plane mechanically. The performance of a modern airplane has made that impossible for even the most gifted pilots. Today, a pilot's movements on the joystick and other controls are sensed by computers, and the computers transform those desires into the electronic signals that control the airplane. In short, the pilot flies a computer, and the computer flies the airplane.

A similar change is underway for cars. An automobile used to be a mechanical product. The cost of the electronics in a car was around 5 percent of the production cost for a 1994 model. It is expected to reach 20 percent within a few years, making a car into a combined mechanical and electronic machine, rather than a purely mechanical product. Cars, too, will become "fly by wire," and will include features which now seem amazing but which will become essential and commonplace. For instance, TRW has developed an intelligent airbag. When this is deployed in an accident, the car requests its location from a satellite-based global positioning navigation device, and sends an automatic 911 telephone message asking for help. This is a graphic example of including both information and service with the product. Modern technology and the emphasis on knowledge and service create many opportunities to add profitable service and information to product, using it as a platform for ongoing interactive customer relationships. New opportunities open up every day for knowledgeable entrepreneurs, and for the companies that encourage them.

> A *product* becomes a platform
>
> A *customer* becomes a long-term subscriber
>
> A *supplier* becomes an associate
>
> A *sale* becomes a continuing transaction
>
> A *supplier* provides an enriching total solution
>
> *Reward* comes from customer-perceived market value

TOTAL SOLUTIONS ARE CREATED BY PEOPLE

In the era of the efficient big producers, how can a small manufacturer survive? If competition were based only on the price and function of product, they would be struggling. But the lower cost and increased capability of technology allows even small producers to think in terms of fusion products, with software, information, and service, bundled together with the product.

Which service or information does one offer on the product platform? These are limited only by the imagination and motivation of one's people. The people of TRW developed their intelligent airbag, only because of their desperation to survive. Faced by pressure to drop prices on their products every year, TRW and other companies are pushed to develop ever more inventive fusion products, and charge for them accordingly.

The opportunities to expand by using the product as a platform for imaginative total fused solutions of product, information, and service are not invented by managers or by designers in a design studio. They develop from the interaction of a knowledgeable and motivated total workforce, working intimately with both customers and suppliers. The solutions are born in interacting work processes within and between interprises, in which people work in self-managing teams in an adaptable organization. Cooperation within and between interprises, and using the product as a platform for services of all kinds, are two sides of the same coin.

SUMMARY

⇨ Competition has moved companies beyond the business of supplying product alone.

⇨ Products and services are becoming platforms for information, and additional technology and service.

⇨ Products and services have changed from being a goal in themselves to being a means to establish close, long-term, interactive customer relationships.

THINGS TO THINK ABOUT

☐ Do your products allow the possibility of adding information and service?

☐ What are the needs of the customers who buy your products? Do you understand their world well enough to understand needs which they themselves may not see?

☐ Does your company have some customers more important than others? Do you rank the importance of your customers according to the potential they offer you to develop more service and information on the product platform?

☐ Do you proactively add services and information and upgrades to the service or product provided the customer, or do you just push product?

CHAPTER 7

TO EACH
HIS OWN

T he charm of the old-world store or the expensive boutique is in its personalized service. "We have a small amount left of the flavored coffee that you like. Shall I send some over?" "How was dinner last night with the Rockefellers? Did they like the outfit you bought last week?" Personalized, individual attention to each customer is the great hinterland of unexplored opportunity, made possible by modern technology. It is no trivial matter that in a 1994 survey, most CEOs targeted customizing of products as a strategic aim for their company.

Each business customer and each consumer customer sees him- or herself as a special individual, but many suppliers persist in seeing them as an abstract average. They deal with a customer as part of a group, not as an individual person or company with individual needs and considerations. The days of herding customers are gone. In the old precomputer days, everyone had to be treated identically, because the effort of recording and recalling individualized information for many customers was impossibly complex. Treating customers individually could then be done only by special, personal attention. Now we have the information technology necessary to deal with many people individually.

When calling a bank or travel agent, whoever you speak to knows all your personal preferences, read off a computer screen at their workstation, and can relate personally even if they have never met you. This approach is even seen in synthetic situations. When buying gasoline at an automated pump, the panel of the electronic pump control prints a personal greeting: "Good evening, Rosie." It is easy to understand how that happens. Your name is encoded into your credit card, and the pump knows the time of day from its internal clock. Personal recognition gives most people pleasure, and today's computer technology allows personalized attention for everyone. The days of being treated as one of a herd are gone. The mass market is no longer anonymous; it now has a face, yours.

Companies are trying hard to find the balance between storing the data needed to fulfill customer needs, while excluding data which infringes on privacy. Finding this balance is facilitated by interacting with the customer, and letting the customer decide which information she or he wishes to allow to be recorded.

THE HIGH COST OF MASS PRODUCTION

The inefficient cost of the mass marketing system, and its incompatibility with individual needs, is revealed by the following statistics:

- Seventy catalogues are printed each year for each person in the U.S., an enormous expense, not to mention waste of paper.
- In the ready-made clothes market, 50 percent of garments fit people well, 25 percent fit not so well, and 25 percent need to be specially made.
- Thirty percent of printed cloth intended for garments is eventually marked down because it is not used;
- Thirty percent of clothing in stores is eventually marked down because it cannot be sold in time.

Similar statistics can be found in many industries. These numbers show the inefficiency of the mass production system, and the opportunity to be found in personalized solutions.

"CONSUMER" AND "CUSTOMER"

We differentiate between the consumer of the product and the customer. The customer for the pair of made-up jeans is the store, but the consumer is the person who buys from the store. To sell personalized products and services, you need to access the user of the product, either directly or via a store or other distribution channel. Personalized solutions can be valuable to both the user and the store, both the consumer and the customer.

> Levi Strauss introduced "Personal Pair" customized jeans for women in 1994. "Among the things a woman really hates to do, is buy a pair of jeans," says Jim Ansel, director of operations support for the jeans maker. "There are lots of try-ons. It's very frustrating." The Personal Pair jeans, which cost $58 compared with $48 for a conventional pair of Levi's, are sold through twenty Original Levi's stores in North America. A salesperson measures the customer, who may try a few pairs to further narrow down the fit required. The salesperson then uses a touch-screen computer to relay information on the customer's desired size, color, and finish to a company computer that generates a digital pattern, which in turn is transmitted to a Levi's factory in Tennessee. The buyer picks up the custom-fit jeans three weeks later. Levi's experience after a year was that returns on the Personal Pair jeans were much less than for standard sizes, that the demand for the customized jeans was strong, and that the stores providing the customized jeans service attracted many more customers than before. Levi Strauss is becoming an interprise, supporting both their customers' retail processes and consumers' life-style processes.

The personalized jeans product has turned out well for consumers, stores and the jeans producer. The people who presumably suffer are the producers and distributors of mass produced, standard jeans, since they will lose their market share. A business which persists in seeing customers as an average mass will inevitably go downhill. In your business, which side do you prefer to be on? The supplier of mass, all-are-the-same products and services, or the supplier which sees each customer and consumer as an individual?

CUSTOMIZING REQUIRES SLASHED CYCLE TIME TO GAIN COMPETITIVE ADVANTAGE

Paying individual attention to customers would be useless if it took a long time to react to a customer idea. Customization without speed is not competitive.

The Textile and Clothing Technology Corporation [TC]² in the U.S., and other groups elsewhere, are continuing to develop the individual solution in apparel manufacturing.

> [TC]² has developed the Quick Response system, where information on sales from the retail outlet is rapidly rolled down the supply chain even as far as the maker of the thread for the weaver of the cloth. Production for a whole industry sector will be pulled by real consumption, not pushed by forecasts. This has reduced the time for the whole cycle from sixty-six weeks to thirty-six weeks, and they aim to arrive at a three-week cycle time.

> The [TC]² capability to reduce cycle time enables personalized clothes. They have developed a system where a consumer specifies a T shirt from a choice of three sizes, four collar styles, two colors which can be different for each of the two sleeves, and initials in any of ten colors. The shirt is chosen, paid for, and delivered within forty-eight hours. This is the equivalent of maintaining over eight million different stock items!

> The next development is a system where the consumer chooses the cloth and the style wanted for a blouse and skirt. She paints a pattern for the cloth on a computer terminal, and stands in front of a 3D camera system to be measured. Within four days, an individually manufactured product, with its personal pattern and design, will reach the consumer.

Adding the capability of personalized products to the quick response creates an unbeatable combination. [TC]^2 is aiming at a three-week cycle time for personalized products. This would enable the industry not only to save the cost of inventory, but also to impose a significant barrier on competition. The textile business then becomes part of the lifestyle process of the consumer, and changes the work processes along the distribution channel. Standard mass-produced high-quality products become cheap, low margin items.

What [TC]^2 is doing for garments, Ross Operating valves is doing for pneumatic valves, providing a customized solution within three days instead of twelve months. As described in Chapter 5, Ross had been a make-to-catalogue company only. A customized order required a year and $100,000. They now can produce a customized valve within five days, for $3000.

Anyone who has ever designed industrial equipment knows the back-of-the-head worry while waiting months for a special component. What if some detail was not thought right through? What if a change is made to the system to which you need to connect the component during the long period that the component is being made? What if there is some unforeseen interaction between the components and the system, such as overheating or a pressure surge? If forced to make a change to the special order, what unforeseen impact would that have? Having a customized engineering product within a few days gives relief beyond description. Instead of paying $100,000 and waiting twelve nervous months, you order it, try it, change it, and even if you need five iterations, you get a perfect product without worrying about the expense and time.

The printing industry is an example of how technology has changed the focus of a whole industry sector. In the old days, lead letters were bolted to iron or wooden plates which were then placed on ponderous machines, allowing specially trained craftsmen to produce many identical printed items. Today, desktop publishing by a computer network with a color printer, enables a person without technical education to rapidly produce any number of high-quality, personalized documents. Printing used to require long setup times for the linotype, printing, and binding work. This necessarily led to long production runs. Today, one-of-a-kind printing is an everyday occurrence. And once we have personally

printed customized items, personalized advertising follows naturally. Not only is the material printed with the subscriber's individual name, but personalized advertising based on the personal
information is added. "It's your birthday next month, treat yourself
to a vacation planned by our experts!"

USE THE NEW TECHNOLOGY NOW—OR SOMEONE ELSE WILL

> The Wall Street Journal advertises the Personal Journal, as
> "the only newspaper with a circulation of one." Specify
> the stock prices and news subjects you want, and your
> personalized issue of The Wall Street Journal will be deliv-
> ered into your computer each day for $12.95 per month
> or $155.40 per year, less than the $164 per year subscrip-
> tion to the paper-based mass-produced issue.

The capability to make such a newspaper is available to any entre-
preneur. Anyone can subscribe to online news and stock price services, and then make as many tailored newspapers as he or she
wants, and send them off. What is the advantage of The Wall Street
Journal in providing the Personal Journal? The technology is indeed
universally available, but not the unique distribution system, reputation, and market penetration of The Wall Street Journal. If they did
not provide the personalized service, someone else would, developing a competitive distribution and support system which would
eventually undermine the market for their mass-produced, regular
morning newspaper.

News media are increasingly using technology to provide customized solutions.

> In 1996, The New York Times began advertising "Custom
> Publishing," creating a magazine "uniquely tailored to
> your company's marketing goal." They work with the cus-
> tomer to understand the target audience, develop the
> right concept, then assemble the best creative team to
> edit and design the magazine. They will promote, sell,
> and distribute the magazine. Their customers include

IBM, USAir, Four Seasons Hotels and Resorts; as well as various state development authorities.

Personalized service is rapidly penetrating service industries, such as banks, in many countries.

> Tom is a teller at the First Direct Bank in Leeds, England, a bank which has no branches and deals with customers only by phone or computer. He enters a customer's name and security password into a computer as he talks to her through his headset. Immediately her balance and account information appear on the screen. This information includes the message "No Adverse Risk," indicating that the bank would be glad to loan her money, together with information that she is a thirty-year-old single mother with a job as a project manager. A listing of all previous contacts with her, including a previous call in which she expressed interest in taking the bank's credit card, scrolls onto Tom's screen, together with a note that information about the card had been mailed six days before the current call. Tom had never seen the customer, and never will, but she received detailed personal attention, thanks to the computerized customer information system.

A generation ago, baking ovens were large and expensive, and only large baking organizations could equip themselves to provide baked goods. Using the currently available technology of small, computer-controlled, efficient ovens, food stores, supermarkets, and restaurants around the world have rapidly taken up the use of small ovens to bake fresh breads and cakes. In doing so, the variety of baked goods available to consumers has increased significantly. Also, because many more people than before are now commercial bakers, many try their personal skills at producing ever-improving goods. Technology is an empowerment tool. The technology of baking has enabled many people, who before had no access to commercial baking, to use their skills. Customization penetrates all businesses, whether service or manufacturing, or small or large.

Mack Trucks took 3,654 orders for CH models in 1995. More than 2,000 of those were for one truck only. Eighty-two percent were for three or less. A customer can choose from fifteen rear axles, twelve engines, thirty-five transmissions, and 4,500 colors. Regardless of how ordinary a Mack truck may look from the outside, it is actually a highly customized business tool built to the exact specifications of very demanding customers. Customers are able to order a truck perfectly tuned to the route they will be driving. Before the company settles on a specification for a customer, Mack personnel take a vehicle and drive it through the actual route the customer will use, noting data on terrain, various speed limits, grades on hills, and the number of stops and starts. They then equip the truck with the collection of components custom fit to maximize operating efficiency and reduce operating cost in those routes.

The personalized solution follows the pervasive penetration of technology into society, and the individual empowerment that follows from it. In every field, systems are becoming available which empower anyone to do work which previously only capital-intensive factories could deploy. Computer graphic systems that design cloth and garments, and computer-controlled cutting and sewing machines, enable small-scale, personalized industry which previously was unthinkable. In the era of the mega-discount store and mega-manufacturer, how can a small business survive? If competition were based on price only, it could not. Customization is a positive survival and expansion strategy for small businesses as well as large ones.

Many companies are determinedly making the change to providing individualized solutions. This will require that they become interprises, trusting people in empowered teams, migrating to adaptable internal structures, changing the relations with their suppliers, and changing the success measures and rewards of their employees—in short, turning the old company structure inside out.

The copy in the ad reads:

"Call your toughest shot. An extra wide bed in the sleeper with horses enough to climb a mountain? Or a lightweight medium-length conventional? Just size up your needs and we'll put together a truck that gets it done. The perfect tool, custom-made and efficient. Whatever combination it takes. The idea here is needs first. Trucks second. It's a way of specing that makes sure the machine you get is exactly the machine you need. A genuine Mack philosophy. No matter how tricky the shot. To find out more or find your nearest dealer, call us at 1-800-922-MACK. Drive One & You'll Know™."

Figure 7.1 Mack Trucks, Inc. Image of Themselves.

Just as there is a proliferation of computer-controlled mixing and baking, or computerized banking, and printing, so you will find that computerized machines enable individualized solutions in almost every activity you can think of. Computer power to make personalized solutions is now available. Use it. If you don't, someone else will.

SUMMARY POINTS

⇨ Each customer can now be treated as a unique individual. Technology enables customization of products and delivery systems.

⇨ Customization is a new frontier of opportunity.

⇨ The technology of customization will not succeed in the old-style, rigid company structure.

⇨ To benefit from the sale of customized products and services, a company must become adaptive.

⇨ Technology empowers people to creatively offer individualized solutions.

THINGS TO THINK ABOUT

☐ Is your information system built around products and orders, or around customers?

☐ Is information relative to everything about a customer concentrated in one system, or is it scattered around the company?

☐ Does the pay and performance system in your company encourage people to concentrate on getting work finished, or does it encourage a focusing on customers?

☐ Do customers get in your way, or are they seen as an asset?

CHAPTER 8

DIVIDE TO
CONQUER

W hat does the interprise do when the market for the product family is crowded and margins are low? If it has direct access to the users of a product, it interacts with customers to offer customized solutions, as do Ross Operating Valves, *The Wall Street Journal*, Levi's, and many others described in Chapter 7. Customization could lead you to a personal solution for each customer, yet you may not have access to the individual customer. How can you compete if you have no direct contact with the product end-user because distribution channels isolate your interprise from direct contact with each consumer?

Where customizing tailors a product or service specially for one customer who actually participates in the process, "sneakerization," or fragmentation divides customers or consumers into many groups, without their personal participation, and provides a different product or service for each group. Nike made this approach famous and their product gave the concept its name.

> Before Nike performed a modern business miracle, sneakers were a $10 product made of canvas tops and rubber soles, a universal solution to the sports shoe need. The conventional approach to competition would have been

to produce a more functional, higher quality shoe for less money. Nike, however, managed to turn a $10 sport shoe into a sneaker priced at over $100, ten times the old price. They did so by inventing a broad range of needs, and the corresponding products. Now, instead of needing one pair of sneakers, each consumer needs several pairs. Nike convinced consumers that since each shoe was advertised for a specialized use, they should pay ten times the old price. There are over 399 types of Nikes: different shoes for running, walking, cross-country running, aerobics, tennis on clay courts, tennis on grass courts, basketball, and many others. The consumer who had previously invested $10 in sneakers, may now have four pairs or more, with a possible outlay of $400.

The cost of production, of course, did not go up by anything near a factor of ten, from $10 to $100. Where most businesses would have competed by reducing costs and improving function of the shoe, Nike succeeded in fragmenting the shoe market, entering the lifestyle process of the consumer, and managing to extract much more money from each of us.

The Nike example shows how the interprise can use fragmentation of its products and offerings as a powerful tool to create entirely new dimensions of competition, and to increase market share in those situations where the customer is not accessible for a customized solution.

"SNEAKERIZING" MATURE PRODUCTS

Experience shows that the life cycle of a product has several stages: first, the novel phase; then rapid expansion; maturity; finally, the phase of decay. These are shown in the figure below. Fragmentation, or sneakerization, can give renewed vigor to a mature product. It effectively changes a mature product into a new product, as shown in the figure.

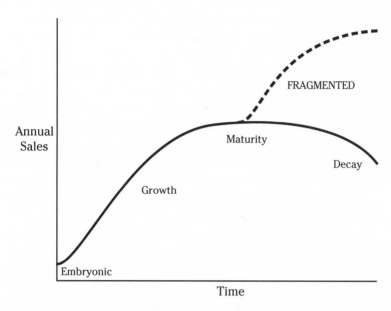

Figure 8.1 Fragmentation and the Product Life Cycle.

What Nike did to sport shoes, Swatch did to watches.

> Swatch redefined a wristwatch from being a time piece, to
> being a playful work of art. If a watch were needed to tell
> time, a person would need only one, but there is no limit
> to the number of decorative works you may want. Redefi-
> nition of the watch as popular art opened the market up
> way beyond the one-watch-per-person idea.

Bicycles used to come in three varieties: regular, ladies', and
sport models. The bicycle market has fragmented, and there are
now dozens of bicycles to suit every circumstance, whether it be
getting around, track racing, road racing, cross country cycling, or
riding through the park. As bicycles were made more specialized,
they have become much more expensive than before, from less
than $100 to many hundreds and even thousands of dollars.

Fragmentation of a product is achieved not only by reconfiguring the product itself, but by paying attention to the information and distribution of the product. The ability to sneakerize is enhanced by seeing the product as a fused whole of hardware, information, and service. Product includes packaging. Supermarkets and other store outlets are increasingly demanding specialized packaging for special events such as sales, or with advertising aimed at the local community. Even though the content of the packaging, the soap, the soft drink, or the candy, may be the same, the assembly of content plus packaging has become fragmented, aimed at providing more variety and choice to consumers. This, in turn, has required that manufacturers and distributors rearrange their operations so as to deal with short production runs of specially packaged material that can be ordered only a short time before they are to be delivered.

Retail products often have special packaging and model numbers, allowing the store to advertise "We will not be undercut," since no one else will ever have that model number. Competitors may have the same product, but with a different model number and possibly different cosmetic details. This requires special packaging, and possibly special assembly and distribution processes. Packaging of product has become a manifestation of sneakerization.

Even as mature a product as car tires can be "sneakerized."

> Goodyear has moved from manufacturing six types of automobile tires, to one type per model year. Now there are many types of tire, and the consumer needs to select just the right one for the model and year of car. The tire market has fragmented. This was achieved by more cooperation with suppliers, by a change to team based efforts inside the company, and by intense interactive work between the people at Goodyear and at its customers. Goodyear has become an interprise.

A decade ago the beer market belonged to the big breweries: Anheuser Busch, Coors, and other large corporations. As the technology of beer making became cheaper and more accessible, small local breweries started popping up, often selling in restricted local areas. At first the large breweries took no notice, thinking that the challenge was too small to be significant. As time passed, however, it became clear that the public liked the niche beers made by local breweries. The big breweries are now creating niche beers too. The beer market has fragmented.

"FRAGMENTING" PRODUCTS EXPANDS MARKETS

People have always liked to be chosen, put above the crowd, honored, made to feel special. In the past, convenience and manifestations of status, such as tailor-made clothes, were expensive and only the rich could afford them. Today's technology enables us to offer a degree of convenience and status to everyone. Fragmenting the product offering into many categories allows you to establish a different product or product configuration for every one of many different categories of consumers. It is a mechanism for making consumers feel a part of a select group, by making the product part of their life process. The user pays for the status it confers, not for the intrinsic properties of the product. If one needed shoes only to run comfortably, there would be no reason to pay a hundred dollars for a pair. In some circles, sports shoes now have a value beyond money. They have been killed for—not because the killer needed a shoe to cover his feet, but because of the self-esteem or reputation that follows from being seen with the shoe. In today's world, the primary use of a Swatch wristwatch is not to tell time. It is used to make a fashion statement, and only secondarily is it used to show the time.

These companies and others have found that they can break ahead of the competition by establishing differentiation by fragmentation. The technique is as valid for commercial and industrial products as it is for end-user consumer products. Competition is pushing car and truck manufacturers to offer an increasing number of options. The few standard configurations of cars made available by each manufacturer have multiplied. As the amount of electronics in a vehicle increases, many of these options can be implemented as software in digital electronic systems. Buyers can now select the give of the suspension, program the car to remember the preferred seat location and radio station for each driver, and more.

Fragmenting product offerings is a technique equally applicable to gaining market share for commercial industrial products. One of the difficult situations faced by a product designer is to match the design of the total product to the limitations of the available components. Product designers become accustomed to the notion that when choosing components, such as a motor, or electronic resistor, they have to search for a component or subassembly with properties close to those needed, then arrange the design around the chosen component. This leads to compromise and tradeoff between different components and parts of the system. If a wide range of component properties is available, and there consequently is only a small difference between the properties of each, the choice of component is much more precise, and the trade-off considerations of the designer much easier. The manufacturer of a large, fragmented range of components, with little difference between each model, establishes a definite advantage in the market.

SIMPLIFY BEFORE SNEAKERIZING

> In a 1995 presentation, Jack Welch, CEO of General Electric, told his senior management to simplify. Simplify to improve quality, simplify to improve interaction with customers, simplify to reduce costs, simplify to expand.

A critical problem when fragmenting product offerings is to avoid being caught up in the complexity trap. Complexity confuses your customer, your suppliers, and your own workers. To provide "sneakerized" products, first simplify both the list of components which make up the product, and the work processes all through the supply chain. Simplification of components and of processes is a necessary precursor to providing both sneakerized and customized products. Simplification of the production processes seeks to reduce the number of components, so you can plan at which points along the supply chain to hold semifinished product. This enables quick response to changing demand for specially designed products, and quick implementation of design variations.

When a new owner bought an ailing Wisconsin scientific instrument company in the late 1980s, he insisted that each new instrument model have half the number of parts of the previous model, and that manufacturing time for the model be cut by 50 to 70 percent. If these standards could not be met, the new model was scratched. This saved not only manufacturing costs, but also the cost of maintenance, service, and spare parts. The policy also led the 650-person company to closer interactions with customers and suppliers, and, to a healthy 18 percent annual growth rate in a business sector that was shrinking at a 2 percent annual rate.

Swatch succeeded not only by redefining the use of a watch by the consumer, but by taking great pains to redesign the internals of the watch, reducing the number of components and simplifying the product.

The interprise makes an effort to simplify matters for the customers. The spread of sneakerized products and services, from running shoes to industrial components, and from banking to insurance services, has created a new type of job. To choose among the confusing collection of ever-changing choices, people turn to specialists to help them navigate the perils of choice. Astute interprises make an effort to help people make those choices easily.

Gateway, which has grown since 1985 to a $2.7 billion company with 6 percent of the market share of personal computers, has worked hard to simplify sets of components and systems of work. They developed a customer interface information system that provides reliable, up-to-date computer configuration data that can be used by all employees, from manufacturing to sales and support. This information system includes lists of components in stock or more readily available, so that demand for less available components can be curbed, and customers can be directed to choosing a computer configuration which can be built and shipped within five days. The data created at the point of sale are instantly transmitted throughout the Gateway organization, to their suppliers of components, and to the service providers, including the shipper UPS. A key to their success has been the simplification of their processes. It is interesting to note that Gateway defines their core competency as a process, in their case order fulfillment, rather than a product. If the computer market eventually decays, they will be well-placed to move on to rapid delivery of other custom-assembled products.

Modern competition is for mindshare. People want sneakerized products, but they do not want the strain of choosing them. In times past, having a better product ensured success, but no more. To sell today, one needs to capture the mindshare of a customer, and in that fleeting moment of attention, seal the deal. Sneakerization implies not only making available many options, but presenting them in a way that will make choosing and paying easy.

FRAGMENTATION MOVES THE FOCUS FROM PRODUCT TO PROCESS

When fragmenting markets and offerings, management attention is redirected from products to processes in the organization. These include product design and production, processes invoked by consumers and customers who use the products, and organizational processes which support the manufacture and supply of products. Traditionally, strategic and tactical planning in companies have concentrated on defining the products to be offered, with little attention paid to the processes whereby those products are defined and delivered. Today, companies pay detailed attention to the processes because otherwise they cannot plan product fragmentation. When dealing with those processes, the guiding principle is to simplify, then simplify again.

In fragmenting the market, do not stop at the first sign of success, but be sure to follow through. After leading the way, Nike did not systematically carry through its market fragmentation. Because of this, Reebok was able to take over the neglected women's shoe market. Nike now acknowledges their failure.

Fragmentation of product and services can change markets and increase profits, as Nike, Swatch, Goodyear, and many others have discovered. It is now an established technique which can be summarized:

- Develop a detailed understanding of how the product or service supports the business or lifestyle processes of the user.
- Design the level and type of fragmentation so as to make those processes convenient, rewarding, and desirable for the user.
- Simplify the design and number of components of the product.
- Simplify the product realization and order fulfillment processes.
- Plan product development and distribution to suit the fragmented market.
- Do not slow down after initial successes, but follow through systematically.

Successful fragmentation requires rethinking one's products and breaking out of the conceptual product-focused box into the world of business work processes and consumers' lifestyles. This, in turn, requires focusing on processes to rebuild the company as an interprise. Fragmentation is engulfing sector after sector, and provides yet another reason that process-focused interprises succeed where product-focused companies stagnate.

SUMMARY POINTS

⇨ As a product matures, it may offer an opportunity for market fragmentation.

⇨ Splitting the market into many small segments can gain you a larger share and margin, but introduces system complexity.

⇨ To deal with complexity, simplify the product, then simplify again.

⇨ Finding profit and opportunity in fragmentation requires interacting closely with customers.

THINGS TO THINK ABOUT

☐ Do you know the lifestyles or work processes of your customers and their customers well enough to design fragmentation of your products or services?

☐ Are your work processes simple enough to permit the move to fragmentation of your offerings?

☐ Do all your people understand your customers' worlds, or do they understand only your company's internal work processes?

CHAPTER 9

GLOBAL
CONNECTIONS

A century ago, businesses sold products wherever traveling salesmen could find doors to knock on. The development of mail-order enabled companies to extend their reach to wherever letters were delivered. Electronic communication has now expanded the arena within which the interprise interacts with customers and suppliers. The space an interprise occupies stretches from the local community to far flung corners of the globe. This new potential is rapidly being taken advantage of by interprises, small and large.

Becoming globally aware means becoming mindful of the sensitivities and customs of each local community in which you do business.

> Lucent Technology, the newly-formed company that split off from AT&T, spent millions of dollars on global research not for their product, but for the name of their company. They needed a name that was not yet registered in any country, and that did not have any unfortunate verbal associations in any language. In their case, doing business globally meant thinking locally everywhere.

As the Internet overcomes its initial teething problems, it is becoming a universal medium of business interaction. Companies are using it internally for its Intranet capabilities, and externally to interact with the world at large. As international travel and electronic communication by telephone, fax, and computer all become more widespread, commercial business transactions reach beyond the old geographical limits.

> The Iscar Company is an interprise, a forward-looking manufacturer of tools such as lathe, mill, and drill bits, located in the Galilee province in Israel. They make use of the new global communication technology to save inventory, reduce costs and speed up the response to customers' orders. The tools are small, so the value per unit volume is high, and air shipment is the preferred method of dispatch. Orders to the value of $1.5 million a day flow by computer network to their headquarters, from all over the world. The orders are assembled the day they arrive and prepared for shipment by an automated warehouse facility. Trucks leave for the airport daily with that day's orders. When each order is assembled, the data is dispatched by network to the customs authorities for automated customs work, and to the agent or branch from which the order originated, so that they can prepare the incoming shipping and dispatching work at their end. Centralization minimizes the need to keep stocks at intermediate distribution points.
>
> Iscar uses international communications for customized items as well as standard ones. They detected a need for quickly delivered, customized, specially shaped mill tools. When work comes into a machine shop, the foreman or owner may discover the need for a specially-shaped tool. Time pressures do not allow him to wait; the metal needs to be cut then and there, and he will improvise with standard tools if a special tool is not available. Iscar, understanding the need not just to sell product, but to become part of their customers' work processes, de-

veloped a design package that runs on a PC, which they give to dealers and customers. The customer designs the tool wanted on the computer in about ten minutes, specifying material type, number of flutes, dimensions, the name to be engraved on it, and all necessary detail. After the design is completed the computer program gives an immediate price quotation for various order quantities. The customer enters a quantity of tools to order, and the computer sends a fax message with all the details to the factory in Galilee. There the data is fed to a computer program which plans and organizes the production. The tool is manufactured within two and a half hours, and dispatched on overnight delivery to anywhere in the world.

GLOBAL PRODUCT SERVICE AND MAINTENANCE

The pervasive penetration of communication is beginning to revolutionize the service of products. It is now commonplace technology to have a machine designed with sensors which monitor performance, sending the information to a control center by wired or wireless communication. The maintenance or repair is planned at the control center. If the repair can be done by software, it is executed remotely, without sending a technician to the machine. This is opening up vast new business opportunities in the service industry, which entrepreneurs around the world are taking advantage of.

Deutsche Telecom has made public the information that they plan and manage the data system for Lufthansa aircraft. Sensors check the functioning of the engines of the airplanes in the air, and send the data via satellite to Lufthansa's control center to be monitored. The maintenance and scheduling of the airplanes is now planned on the actual state of the engines, not on the basis of hours of use as in the past. Traditionally, machine maintenance is scheduled after a certain number of hours of operation, so the engine is often worked on before it becomes necessary. Remote monitoring leads to maintenance when it is required, and not before which saves Lufthansa money.

Ship owners use a similar system. The engines are monitored while the ship is at sea, and the data relayed to a
control center which plans operation and maintenance.

This new, emerging service economy provides vast new opportunities for the interprise. Machines, whether cars or household gad-
gets or industrial machines, have sensors built into them. These are
then connected to a worldwide communications web. This allows
constant monitoring of the performance and status of the machine,
so that planning of operations, maintenance, and repair is scheduled according to the actual condition of the machine, rather than
the hours it has worked. '

Owning a machine can be a stressful experience. The owner of a
car or washing machine constantly worries about rattles and
noises. If the machine needs repair, he or she is concerned about
whether the correct repair is being done, and whether the cost is
excessive. Sensors in a machine read the status of its various components and systems, and, connecting those sensors to a communication system, distributes the data to anywhere in the world. The
technology is the same as performing a medical test on someone
then connecting a specialist doctor remotely to get her or his diagnosis. These developments in service technology will eliminate the
monopoly of the local repair service, because the data can be read
electronically by anyone, anywhere in the world. The change to sen-
sor-based and computer-network-based service creates a new and
easier lifestyle for the owner and gives the interprise myriad opportunities to become part of a customer's business or lifestyle
processes. These new opportunities include sensor data monitoring, repair work, innovative insurance policies for those owners
connected to the data network, brokerage to find repair experts at
remote locations, and component expediting, so that spare parts
reach any destination in a short time. The networked infrastructure
for such servicing is being dealt with by CommerceNet of Palo Alto,
California.

GLOBAL INFORMATION INTERPRISES

Banking is increasingly, and understandably, a communication-based service. In the old days, when money was all gold or printed paper, the primary purpose of the bank was to hold your gold, paper, or precious jewels. Today a bank holds few physical valuables, most of its holdings are bits of electricity in computers, which are easily stored and shipped. That ease of shipping and storing data is changing banks into interprises as they rethink their role, using communication to become part of their customers' financial processes.

In going online, a bank easily steps beyond the boundary of a local area and is able to interact with customers wherever electronic communication exists.

> Hamilton Bancshares of Columbus, Ohio, is replacing 40 percent of its branches with small, full-function, remotely-staffed banks. They and many other banks everywhere, have discovered that the bulk of their profits come from young and middle-aged customers who are frequent users of ATMs. They, therefore, establish workerless banks, branches of the interprise, that provide twenty-four hour interactive video access to remotely located personnel for transactions that require it. Most transactions are handled on enhanced ATM machines.

> The First Direct Bank in Leeds, England, has half a million customers and not a single branch. First Direct is Britain's fastest growing bank, one of the world's leading telephone-only banking interprises. Its customers call at all hours to pay bills, invest, and arrange mortgages. "In Britain the statistics show that people are more likely to change their wife than change their bank," said Peter Simpson, First Direct's commercial director. Yet their service, which is said to be distinctively better than the local branch of a conventional bank, gives them "10,000 new accounts a month. That's like two or three new branches a month." Thus, the company can grow, without the ex-

pense and trouble of finding and refurbishing buildings and hiring local staff. The lure of branchless banking, which gives better service at lower cost, is changing arm's length banking into an interprising operation.

In late 1995, a company called Electronic Share Information (ESI) Ltd (http://www.esi.co.uk/), a member of the London Stock Exchange, located in Cambridge, England, started offering their service online. Their customers can receive information about investments and market analysis, as well as trade in stocks and securities at much cheaper rates than in the old-style companies. Through these convenient computer-networked processes, the customer can directly watch and analyze financial markets, rather than depend on the reactions of an arm's length broker. When they have enough subscribers, ESI will also be able to gauge public reaction to the market by monitoring the items people browse and use on the network. By doing this without recording who initiates each individual interaction, the information can be gathered confidentially. They are a financial interprise on the Internet.

THE BATTLE FOR MINDSHARE

Finding and distributing the right information to the right user, when and where needed, had in the past presented a huge challenge. With the explosive growth of communication, distribution is no longer a problem. For the sender, the problem is now how to get the recipient to pay attention to the sender's piece of information—to find his or her needle in the information haystack. Conversely, the recipient is faced with the problem of filtering information, so that his or her attention can be efficiently focused.

> **When information is plentiful, information about information becomes as valuable as the information itself.**

Competition used to be for distribution of information; now it is for mindshare. How do you get a potential customer to pay attention for a fleeting minute, and, in that single minute, get your message across and close the deal? The soundbite is here to stay, not only in television news, but in business interactions. Getting and keeping customer attention has become a significant aim for the interprise.

In the past, the most intense interaction that companies had with customers was when the customer had a complaint. That is if they were lucky: most complaining customers complain only to other customers, and the company never hears from them. Limiting interactions to noting customer complaints however, is a distorted mode of communication. To become part of a customer's lifestyle or work processes, you need to find a way to a more meaningful dialogue. Ford Motor Company found a way to do this with its Taurus model by using an advertising agency on the cutting edge.

J. Walter Thompson is showing the way into a new world of marketing in the global, electronically connected world. Ford turned to the agency in 1990 to organize direct electronic interaction between consumers and the people who design their cars. The advertising company organized a new division in Detroit, called J. Walter Thompson/OnLine. They sent notification of online conferences to owners of Taurus cars, and at prearranged two-hour sessions, arranged for the Ford Taurus designers to respond online. In addition, an e-mail address was provided to the public for messages and questions of any kind about the car. About 300 consumers participated in the first interactive sessions, and the e-mail address initially received about 7000 messages a month. At the free 800 telephone number, about 90 percent of the calls were complaints, a statistic common for phone calls. In the e-mail system, only 35 percent of messages were complaints. The rest were all manner of questions and ideas about the Taurus.

The online system not only provides designers with valuable consumer feedback, it enables consumers to become part of the product design and realization process, leading to buy-in and "ownership" of the product. In the modern world of innumerable quality products, competition between interprises is increasingly for mind-share. Ford, aided by J. Walter Thompson, is using online abilities to establish a foothold in the mindshare of a part of the consumer market. Ford estimates that each one point gain in owner loyalty could be worth $100 million in profit. These are big stakes.

Small businesses as well as large are engaged in the battle for mindshare, and communication empowers small organizations, giving them new opportunities in the global marketplace.

> Since the fall of 1995, an interprise called e-Coupons at the University of Ann Arbor in Michigan (http://www.e-coupons.com), offers lists of local merchants with advertising. Clicking on a store name results in a printable coupon from the computer screen. The stores get advertising, demographic data, and surveys about products and services. They get an insight into the lifestyle processes and hence the needs of their customer base. The students get advertising information and discounts.

> Federated Flowers has become the largest floral products company in the world by developing into an interprise. In the late 1980s they could have simply continued to provide their free telephone ordering service for florists, but their far-sighted executives understood that with developments in communication technology, others could easily do the same, robbing them of competitive advantage. So they created a constantly updated electronic catalog of floral arrangements, designed for various occasions, ages, ethnic groups, and geographic locations. In addition to making these products available online to over 35,000 florists, Federated injected themselves into their suppliers' work processes to ensure quality and reliability of the operation. Federated takes responsibility for quality to the customer, and operates a self-funded 100 percent guarantee of service: payment for any floral arrangement that does not meet the expectations of both requester and recipient is quickly reimbursed together with a small gift and apology to both parties.

Federated Flowers has entered the lifestyle of consumers, as convenient providers of floral products and
services, and has integrated its operations with the work
processes of its suppliers. It offers a popular online input
device in which a customer enters all the "date, name,
event, remarks" information for occasions about which
they wish to be reminded. Before each date, an e-mail re-
minder message is delivered to the customer, who may
place an order. Federated Flowers is an interprise with no
inventory, no shops, and no product brands. It exists in
cyberspace.

In the old-style enterprise, distribution of information was a
problem, and the measure of efficiency of advertising was cost per
thousand, where the thousand represents a thousand people estimated to be watching a television screen or reading a magazine.
This does not measure the effectiveness of the advertising—it measures the effectiveness of the distribution of the television image or
of the magazine. The move to interactive information distribution
methods, such as the World Wide Web, enables the interprise to
measure the true effectiveness of advertising. On Web sites, it is
now standard to monitor the number of "hits" on the page, who
made them, and when and how much time is spent in exactly which
parts of the site. The advertiser gets specific feedback on the attractiveness of the information, some data on who accesses their
site, and which information attracts their attention.

MULTIMEDIA FOR MARKETING

Harley Davidson's penetration into the lifestyle of their customers
is extraordinary. Owning a Harley has become a motorcycle cult.
Bikers meet often, exchange experiences, and show a loyalty to
Harley Davidson which borders on the fanatic.

Harley-Davidson has an enviable problem. Their motor-
bike is so popular that dealers cannot keep a demonstra-
tion machine in the shop—it is immediately sold to an
anxious waiting customer. As a result, dialogue with a potential buyer about options can be difficult. To solve this

problem Andersen Consulting created a multimedia computerized system that can be used in any showroom. The potential customer selects one component option, sees the bike with the option from different spatial directions, hears the sound in stereo, and gets the price for their selection. They continue to select options, comparing the sound, sight and price until a customized configuration is chosen.

Similarly, Chrysler, working with the Texas software company Trilogy, has combined World Wide Web and multimedia technology, to provide a system whereby anyone can customize and order their own car. This not only enables the consumer to feel ownership of the car they designed and ordered, but keeps note of which options are tried and selected, giving valuable information to the designers about the popularity of various options.

The mundane world of billboard advertising was propelled into the modern era by James P. Andrew, the president and CEO of Matthew Outdoor Advertising Inc., a little known but pioneering enterprise.

Matthew Outdoor Advertising Inc. of Bangor, Pennsylvania, is a company of fifty people, selling billboard space in over 70 percent of the world's media markets, with around 2,000 displays. Selling advertising space at a remote location is difficult. The buyer wants to know exactly what he or she will be getting, and that is not feasible without an expensive visit to a distant location. It was not feasible, that is, until Matthew Outdoor formed an alliance in 1993 with a British company, Key Systems.

Key Systems provided computer software which could accurately simulate a billboard, along with maps and statistical data of the area. Matthew Outdoor included in the software, video clips of each of their billboards filmed through the windshield of a moving car. When making a presentation to a potential buyer, artwork for the client's company is inserted on the simulated billboard. The exact location of each billboard is established from a global positioning satellite system; this connects the image with a database containing specific economic and demographic data for each billboard location. Clients are shown hard numbers, such as distances to retailers and demographic data which allow them to make detailed and informed decisions about the utilization of their advertising budgets.

When hard-nosed financial organizations move into multimedia for routine and mundane efforts, we know that multimedia technology has penetrated into society.

The nearly 200,000 shareholders of the First Union found a CD-ROM tucked into the cover of the 1995 annual report of the Charlotte, North Carolina bank. The computer disk is not just an electronic copy of the paper report. It includes video and animation and has a browser that links investors to the bank's home page on the World Wide Web. Says Edward Crutchfield, First Union's chairman and executive officer, "Investors don't have to plow through sixty pages of text if they just want to find what the president said about the credit card business." Shareholders also can create customized charts of the performance of the bank's stock. The shareholder report has advanced from being a linear report on paper to being an interactive, multimedia hypertext document.

In 1995, only about six companies published CD-ROM versions of their annual reports. The number is continually growing. CD-ROM interactive documents are rapidly becoming a way of attracting attention. In the interprise, which depends on mindshare to compete, using CD-ROM multimedia technology for electronic documents will soon become routine.

SUPPLIERSHIP, BOTH GLOBAL AND LOCAL

Communication connects the supplier to the customer directly, whether the supplier is around the corner or around the world.

> Communication has been a part of the sales plan for the prisoners at Oregon State prison. The inmates—drug offenders, murderers, and rapists—make a line of clothing called "Prison Blues." Advertising slogans include "Made on the inside to be worn on the outside." The jeans, prison jackets, and caps are made for the Oregon and South Carolina prison system, but are also sold at stores, by mail order, and on the Internet. In the case of this particular factory, the workers cannot travel personally to their customers, so they use the Internet to connect to them.

The global revolution in suppliership has also spread beyond legitimate business. Car theft, for example, is an international business. U.S. Customs Service agents and insurance industry investigators estimate that about 300,000 of the 1.5 million U.S. cars stolen each year end up in foreign countries. Joe Pierron, head of international investigations for the National Insurance Crime Bureau says, "The thieves are learning to do the same kinds of sophisticated analysis of investment risks and cost benefits that large legitimate corporations rely on." International communication and access to information and computers is becoming universal.

THE GLOBAL SCOPE OF LOGISTICS

Communication is not limited to electronic interaction. Global and regional movement of people and goods, by truck, train, ship, and airplane, enhance and intensify the global movement of product.

> The downsizing of military forces around the world has lead to an opportunity to facilitate the logistics of international trade. Local and national governments, working with investors, are converting military airbases to cargo transportation hubs, since the bases are often situated where road, rail, and sea traffic converge. Such global

transpark sites are already, or soon will be, established in Hong Kong; Subic Bay in the Philippines; Thailand; Munich, Germany; and Texas and North Carolina in the USA. A twenty-four-hour cycle time transportation service is being planned between any pair of hubs. The time is measured from just before the shipment enters the sending airport to the time it has left the receiving airport. This efficient air shipping capability is a powerful enabler of the interprise.

The Caterpillar company is justly famous for their tractors and heavy earthmoving equipment. It is their logistics capability, however, unmatched by any of their competitors, that is a considerable influence in keeping their market share. When a Caterpillar machine breaks, the hourly cost to the contractor using it is high. Caterpillar can distribute 98.5 percent of their 150,000 line items within 24 hours anywhere in the world. Caterpillar does more than sell products; they support their customers' earthmoving work promptly.

The impact of the new communication technology goes far beyond the online adaption of cable TV, or the ease and economy of e-mail. It goes beyond making the old business environment quicker and more efficient, it is changing the environment itself. By altering the way we communicate, technology is challenging businesses to revaluate why and how they are communicating—whether with their stakeholders or competitors, the local community, or the world at large.

SUMMARY POINTS

⇨ The new communications technology enables every company to go global.

⇨ Communications has revolutionized hardware maintenance, as machines around the world are equipped with sensors that allow remote monitoring and diagnosis.

⇨ A new service economy based on communication is arriving.

⇨ Information about information is becoming a key resource.

⇨ Communication technology injects a supplier into its customer's business processes and lifestyle.

THINGS TO THINK ABOUT

☐ What are the multimedia opportunities for your company or business unit to capture your customers' mindshare?

☐ If you work in a unit of a large corporation, is your business unit taking advantage of the Internet to increase your exposure to prospective and existing customers?

☐ Do you know what customers *really* prefer about your products and your company, or do you know only what they tell you? Could you learn more by having customers "play" on a multimedia computer program at choosing options and configuring products and services?

☐ If you work in a local business, should you be thinking of expanding into a whole region using computer networking?

Part 2

MUTUALLY
PROFITABLE
RELATIONSHIPS

CHAPTER 10

MULTIPLE POINTS OF VIEW

In the era of mass production, the belief was that everyone who had the same needs should be treated in the same way. Henry Ford's famous statement "you can have a model 'T' in any color as long as its black," perhaps best epitomizes this point of view. As we enter the era of knowledge-based production, however, one of the most fundamental new capabilities is an ability to treat people individually in a way that is appropriate to each person. This important change in perspective is rapidly becoming a critical success factor for businesses.

Mass production was not merely a way of producing goods and services; it also provided a model for society and societal values. In this model, a common notion of fairness held that to treat people fairly was to treat them identically, a value system that directly reflected the mass production paradigm. In the knowledge-generation paradigm, treating each person as an individual is now possible, and society is redefining its concept of fairness.

If people are treated individually, multiple points of view soon arise, which the interprise must somehow note and respond to individually. For example, computerized order fulfillment systems not only keep track of a customer's order, and what stage in the fulfillment process it has reached, but also record details of the cus-

tomer's preferences, and each interaction the customer has had
with the supplier.

> The age-old business of gambling, that for many decades
> in the U.S. was monopolized by a few interests centered
> in Las Vegas, is becoming more and more competitive as
> many other states legalize gambling in certain areas. An
> inventive company has come up with the "Pittrack" card,
> which looks like a credit card, and replaces the chips that
> are used to make bets. Customers enjoy the card's ease of
> use, casinos like it because it helps them with the ever-
> present problem of security, and banks are interested in
> its potential. The card itself is capable of capturing infor-
> mation—the type of game a customer favors, how often
> he or she plays, how big the bets are—that can be used to
> market customized deals. The company's ingenuity lies
> in the widely varied business interests they were able to
> address with a single product.

THE CUSTOMER IS NOT KING!

When you can take into account a person's point of view, and treat
someone as an individual, it will naturally lead to two important
business phenomena: First, that there are many points of view, in
addition to the customer's; second, by treating people as individuals, you have begun a relationship with them. Relationships between people, between organizations and people, and between
organizations will dramatically alter the way competitive decisions
are made, and play a major role in determining competitive success. These relationships are discussed in the next chapter.

When you give a group of people the same facts, they invariably
interpret them differently according to their different perspectives,
backgrounds, and experience. In the illustration below, a group of
people are looking at a house.

While they are all looking at the same thing, each sees it from his
or her own frame of reference. We call this phenomenon a personal
context. So, for example, the person on crutches focuses on the
steps, while the tax assessor on the right is thinking of the possibility of a huge assessment.

EVERYBODY HAS A POINT-OF-VIEW!

Figure 10.1 Multiple Points of View.

The concept of a personal context is not new, but it becomes more important in the paradigm switch from mass production to the more agile competition of the twenty-first century. The benefit of treating the customer as an individual comes from the idea that each customer will see a company differently based on personal needs and opportunities. What the customer values is based on the context or situation that he or she is operating in, or what some call the customer point of view.

Much of the customer-focused organizational literature takes this idea a step further by suggesting that businesses make the customer point of view king. But the customer cannot be king! In the service of royalty, there is no equal relationship, no free exchange of ideas, no need to understand the king's personal point of view. In a "customer is the king" modality, businesses strive to give exactly what is asked for, instead of establishing a dialogue that leads to a mutual understanding of how to best balance need and capability. When the supplier indeed understands the customer's business, it can do more than even the customer expected, to provide real and measurable value, often with less waste than before, and with greater measurable benefit to both customer and supplier.

In order to truly enrich the customer—to provide it with something that has measurable value—the supplier must know the customer's business as well as its own. The very fact of understanding the customer's business needs, is itself of value. Recently, a panel of CEOs described what it took to be world-class. Five out of the five expressed a need to have their relationships with customers and suppliers enhanced by knowledge of how one might offer significant competitive advantage to the other. Lucent Technologies, for instance, expects its suppliers to know and understand how to make its microelectronics manufacturing capabilities more competitive, and *The New York Times* expects GMA, who specialize in mailing equipment and service, to supply world-class solutions to their needs.

The customer-supplier relationship now depends on an exchange of knowledge over time, with the supplier using its knowledge, information, capabilities, and leading edge technology to help the customer best achieve their competitive goals. "Suppliers must know the customer's business better than the customer," says Agile Web President and CEO Bill Adams. The CEO of Surtech agrees. "We understand the customers so well that they view us as a resource for them. We view our suppliers that way." Frequently, this means educating the customer about potential opportunities.

THE PARTNER PERSPECTIVE

The customer point of view is one of many possible contexts that interprise leaders must deal with in today's changing competitive environment. It is the one that was most universally ignored in the past, and thus is now sometimes overemphasized. There are a large number of points of view, and interprise leaders must examine a broad set of varying perspectives to determine which ones, at any point in time, are likely to have a strategic impact on the interprise.

Partners are increasingly important to the interprise. The customer, satisfied with the supplier's integration into its work processes in one place, may ask the supplier to do the same for them in different locations. The supplier can do this by partnering with locals in each new geographic area. Federated Flowers, the company described in Chapter 9, which provides individualized floral arrangements anywhere, operates entirely through partnership agreements with local florists. Sometimes a customer may need a goods and service combination which the supplier cannot provide. Or a company may simply need something done in a specific time frame. Partnering, especially in a virtual organization relationship, has become a popular way to meet these needs. Apple invited Sony to be their partner in bringing out the Apple PowerBook because they could do it together more quickly than Apple alone. In short, partnering is becoming an increasingly important tool or option in management's bag of tricks today.

OTHER STAKEHOLDERS

Employees, owners, customers, partners, and even society at large also have points of view critical to business success. These groups are often called stakeholders because they all have a stake in the success of the interprise. Each one, however, sees the interprise differently. Because of this, it is important to understand why any one point of view is or is not critical, and the context in which that occurs. Over the course of time, various points of view will rise and fall in priority and importance to the success of an interprise.

Each organization can identify a large number of points of view, which may or may not have a major impact on the organization.

- Customers
- Suppliers
- Owners and Investors
- Senior Management Individuals
- Employees
- Government officials and Agencies
- Universities and Academics
- Partners
- Competitors
- Board of Directors
- Management
- Society
- Neighbors
- School systems

This list is by no means complete, and the importance of different groups will vary from business to business. In addition, each group may have multiple viewpoints and constituents. For every point of view, the needs and desires will be different. The interprise must understand the point of view of each stakeholder in order to supply value to them, and it must convey a sense of safety, reliability, and stability over time. The interprise has to determine which points of view are relevant, which conflict and so threaten the interprise, and which are different, yet will not cause harm. From this determination will come an interprise mission that further enriches the interprise as a whole by balancing the value provided to each stakeholder.

Over the last few decades, manufacturing firms have been required by government regulation and public pressure to address such issues as employee and community safety, and environmental hazards. Expenditures and procedures that are accepted today as a matter of course would have been considered exceptionally enlightened in the past. Companies have had to respond to the points of view of their stakeholders. Tensions between stockholders, employees, management, and the community have always created problems for business leaders. As the interprise focuses on interactive supplier-customer relationships, and builds relationships within and between other interprises, reconciling the various points of view will require more compromise than in the past. The need to motivate and reward risk-taking teams of employees, and enrich enlightened and demanding investors, while remaining responsive to changing social values, is a formidable task.

We recommend this exercise to many teams of senior managers:

- Identify explicitly all of the points of view which are important in determining the success of the interprise in the next few years.
- For each point of view, determine a vision of success and the metrics to be used in measuring success.
- Identify obstacles which must be overcome to achieve the vision and get good metric scores.
- Resolve conflicts in the vision for each point of view.
- Develop a unifying point of view and vision of success, paying careful attention to preserving the vision in the metrics.
- Identify a method by which each obstacle can be overcome.
- Convert all of the above to time lines with a clear link to how they will realize the vision.
- Identify critical success factors required on the time line to achieve vision metrics.

In the past, the various points of view were often considered as an afterthought, or when they elbowed their way to the forefront. It is better, however, to proactively and methodically take them into account, in order to preempt unpleasant surprises.

THE EMPLOYEE PERSPECTIVE

More and more large firms, having downsized, are facing employee morale problems. The old formula for the relationship between employees and the corporation was based on loyalty: the employee dedicated his or her life to the company, and in return he or she expected a lifetime guarantee of employment, family health coverage, and a pension. In the new environment, no company can fulfill these expectations.

Instead, interprises are looking at other ways to reward employee loyalty. They promise to keep education and learning experience at current levels, so employees can transfer their skills should the firm be unable to maintain them. They treat employees as individuals with appropriately structured experiences and rewards. There is more recognition of lifestyle needs—child care centers, laundry pickup, well care, health care, fitness centers, diet programs, and so on. Interprise leaders are working with employee groups to make jobs more rewarding and fun. They are trying to create a community of mutual benefit now and for the future, and to provide real and substantial benefits to replace the broken promise of lifetime employment. The challenge leaders face is to carry this type of innovation to other stakeholder relationships.

> In 1985, the chemical manufacturing giant DuPont did something unusual in order to retain good employees. They surveyed employees about childcare needs and found to their surprise that for half the women and a quarter of the men, managing childcare and the job was so difficult that they had considered leaving for a more flexible company. Cindi Johnson, senior work-life specialist at DuPont's Wilmington, Delaware headquarters, says, "We were stunned." Since then DuPont has had a constantly-evolving work-life program which helps employ-

ees balance the pressures of work with childcare, eldercare, and other family needs. In 1995, DuPont estimated that the value of the resulting increased performance, employee retention, and stress reduction and reduced absenteeism was worth 637 percent of the cost of the program. The key, say DuPont executives, is to tailor the program to each employee's needs, not just offer a good-sounding menu of choices.

The old-style, producer and product-focused business took notice of few points of view—often only considering what the owner might think. The business aim of the interprise is to become part of the customers' business or lifestyle processes. To do that successfully, it understands that not only is each customer's point of view different and constantly evolving, but the points of view of all the stakeholders must be considered and balanced against each other. Fairness and success are achieved not by treating each stakeholder similarly, but by treating each appropriately.

SUMMARY POINTS

⇨ In an interprise, the various points of view of the stakeholders need to be taken into account and reconciled if necessary.

⇨ The idea that fair treatment is equal treatment is no longer true. To treat people fairly is to treat each one appropriately.

⇨ Relationships change constantly, and need ongoing care and maintenance.

THINGS TO THINK ABOUT

- ☐ Who are the stakeholders in your organization?

- ☐ List the points of view of each stakeholder.

- ☐ Prioritize your dealings with different stakeholders. What conditions could cause that prioritization to change?

CHAPTER 11

BALANCING THE RELATIONSHIPS

T he change from selling identical products with identical services for a fixed price in arm's length transactions to interacting with customers to provide specific solutions for varying prices has profound and far-reaching implications. When Ross Operating Valves developed mutually beneficial relationships with their customers, they profitably expanded their business from catalogue products to customized solutions.

> FedEx does more than deliver packages. Their customers can obtain up-to-the-second information as to where their package is by calling an 800 telephone number, by interfacing with the FedEx Web site, or by using a specially designed computer software package. They can also use a caller identification program that saves their customers the inconvenience of spelling out information, such as their address and phone number, over and over again. FedEx, like UPS and other courier companies, now provides total logistics services for business customers. Supporting the relationship is important in nurturing customer loyalty.

In an arm's length transaction it was not possible to sell individualized solutions; the customer simply ordered the required product or service from the supplier, and neither party needed to know much about how the other did business. Under these circumstances, the customer would never become aware of the full range of capabilities that the supplier could provide. Equally, the supplier would not learn the full extent of their customer's needs. In the U.S. defense industry, where government regulation enforced arm's length relationships, development cycles for weapons systems were notoriously slow, often taking as long as fifteen years, by which time the solution was out-of-date. Under the pressure of war, however, the military interacts directly with industrial companies, and solutions are rapidly produced. In business, if a supplier wants to work with a customer to support the customer's profits, supplier and customer must interact closely.

Although competition for the last hundred years has been based on selling identical products at a given price, a growing body of evidence shows that business in the early twenty-first century will be won and lost based on relationships formed across company lines. The familiar and comfortable world of arm's length transactions and fixed prices is being replaced by interactive relationships and flexible pricing. If an interprise has a relationship with a customer in which it provides specialty goods and services that have a measurable impact on the customer's bottom line, it does not make sense for the interprise to charge a fixed price. The supplier should set a fee commensurate with its efforts and the customer's benefit. In these new relationships, each situation will be different and must be treated appropriately.

The Interprise Relationship Model below is a way of mapping the interprise relationships now evolving. The three dimensions define key management decision-making areas in managing the interprise for mutually profitable relationships.

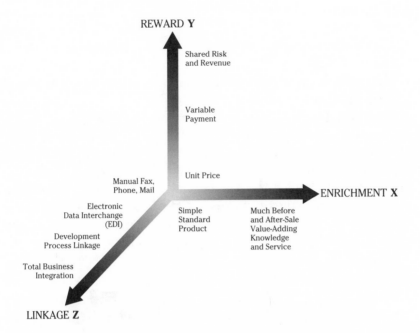

Figure 11.1 The Interprise Relationship Model.

1. The horizontal axis represents the added-value conferred on a customer by a supplier. This can be any combination of hardware, service, or knowledge over the entire lifespan of a project or product. These can include special design, customized delivery, upgrades, maintenance, decomissioning, and recycling.
2. The vertical axis represents the reward or payment made by the customer, from fixed price through variable payments to shared risk and reward.
3. The third axis represents the degree of business linkage between companies, starting with disconnected processes, such as mail, fax, and telephone, through electronic data interchange, to a complete integration of operations.

In short,

- The horizontal axis represents what the *supplier* gives the *customer.*
- The vertical axis represents what the *customer* gives the *supplier.*
- The third axis represents how they *work together.*

These three factors define a space within which a relationship between two companies is located. Figure 11.2 shows the same values using parallel arrows. It is crucial to understand that efforts made in any one of the three axes must be balanced by efforts in the other two. Movement has to be coordinated along all three axes in order to achieve a higher value in one of them.

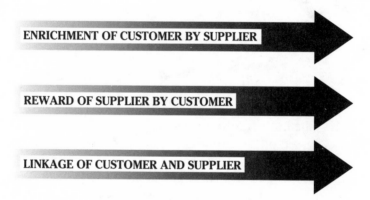

ENRICHMENT OF CUSTOMER BY SUPPLIER

REWARD OF SUPPLIER BY CUSTOMER

LINKAGE OF CUSTOMER AND SUPPLIER

Figure 11.2 The Three Axes of the Interprise Relationship Model.

THE ENRICHMENT DIMENSION

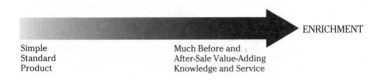

ENRICHMENT

Simple   Much Before and
Standard   After-Sale Value-Adding
Product   Knowledge and Service

Figure 11.3 The Enrichment Dimension.

A supplier who provides a specialized product or service that makes a measurable difference to the customer's bottom line is operating in the enrichment dimension (See Figure 11.3). The supplier will look for opportunities to customize and individualize goods and services to make them more valuable to each customer. Knowing, tracking, and understanding how the customer sees value becomes a critical success factor in relationship building.

> When Landis and Gear, who make control systems, learned that their customers wanted to have energy efficiency and climate control in their office buildings and shopping centers, they did not focus on selling controls and a service contract, but on managing the energy and climate control systems in the buildings and shopping centers. This provided a significant measure of enrichment to their customers who now hire them to manage energy, not supply controls and service.

Generally, a supplier seeks a relationship that will last over time, giving them the opportunity to streamline work processes and integrate systems for greater efficiency and profit. The supplier also integrates their own suppliers into their work processes, creating an enrichment chain that provides value at each level of service or production for both supplier and customer. Wal-Mart does this with their suppliers to provide better value for their customers. In the enrichment dimension, the interprise will even try to measure the benefit provided to the customer's customers so they can better quantify the benefit, or enrichment, to their own customers. Texas Instruments uses the "two up, two down" philosophy, looking "up" to the customer and the customer's customer, and "down" to the supplier and the supplier's supplier—five companies in all.

THE REWARD DIMENSION

In an environment where businesses provide enrichment-based outcomes, fixed prices are being replaced by shared risk and reward systems. This is the reward dimension, in which price follows value.

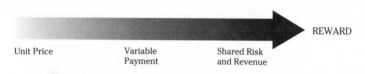

| Unit Price | Variable Payment | Shared Risk and Revenue | REWARD |

Figure 11.4 The Reward Dimension.

When a supplier can effect savings for a customer requiring investment by the customer, it is becoming common for the supplier to make the investment themselves in return for a share in the potential savings. By offering an enriching solution, the supplier has the opportunity to benefit directly from it, while the customer is able to spread its financial risk. This provides a linkage between the enrichment dimension and the reward dimension. It is similar to value-based pricing because of its linkage to the price that the market will tolerate. Many franchise operations use value-based pricing. The buyer pays for the use of the name and expectation in the minds of customers, but his or her fee is based on sales or market experience. Similarly, the short term automobile leases for luxury cars with high residual values is partly derived from the relationship-based thinking of price follows value.

The key to understanding and using this dimension of the relationship model is to abandon the concept of fixed prices under any and all circumstances. There are clear and important situation dependencies to a relationship, which determine not only how much enrichment the customer will receive, but in what way the supplier should share in the risk and the reward. For instance, it is increasingly common for supplier and customer to split the supplier's cost savings. Companies like Harley Davidson share with their suppliers the savings from joint cost-cutting efforts. This gives the suppliers an incentive to cooperate in reducing costs.

THE LINKAGE DIMENSION

The third dimension of the relationship model for the interprise deals with building linkages between the various participants in the interprise. It is a way of quantifying the linkages, and suggesting the types of interaction which may lead to strategic advantage. Movement in any one dimension will often suggest movement on the others, although not always in a linear fashion. For instance, a move to enrich the customer that leads to shared risk and reward will require significant movement along the linkage axis as well. In that relationship, each side has to be trustworthy, which means reliably doing what you say you will do.

The linkage dimension is about linking people, assets, and ideas in order to gain competitive advantage. It is about sharing information in common work processes to make relationships more effective, but it is not only about utilizing electronic information and communications systems. Linkage implies cooperation between people, who build a relationship based on mutual understanding, an evolving sense of trust, shared ethics, and values. The linkage dimension is first and foremost a human dimension about relationships; the use of information sharing systems ensues from this.

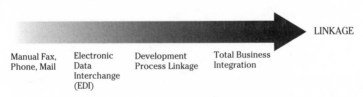

			LINKAGE
Manual Fax, Phone, Mail	Electronic Data Interchange (EDI)	Development Process Linkage	Total Business Integration

Figure 11.5 The Linkage Dimension.

To be useful, a linkage must result in measurable improvements in the relationship between two or more elements of the interprise. For example, the benefits routinely attributed to concurrent engineering are higher quality, decreased cost, and decreased development time. These accrue to an interprise when the organizations working together have agreed to share and use information with each other. E-mail, electronic data interchange (EDI), the Internet, and other information systems may be crucial to this process, but they cannot substitute for people who have found a way to cooperate. Linkage is a means for the relationship to achieve a profitable end.

> JM Mold is a small Ohio company who worked interactively with their customer, Encor, and their customer's customer, Whirlpool, to rapidly create a part needed for Whirlpool's lint-free clothes dryer. JM Mold succeeded because they could sustain the intense interactive relationship this effort required.

The dimensions of the Interprise Relationship Model are a tool to help you position relationships within your interprise. As we have said, the model is systemic in nature, so efforts in all three dimensions must move up together in order for the interprise to move to higher levels in just one of them. A highly linked interprise in which people constantly exchange ideas and information will use a value-based reward approach, as opposed to a fixed price for fixed products. Similarly, the interactive interprise will only function well with a high degree of business linkage to other interprises; an interprise using a price follows value methodology cannot sell goods and services at a fixed price without linkage. Simultaneous movement in all three dimensions is explored in the discussion of virtual relationships in Chapter 16.

SUMMARY POINTS

⇨ The change from selling commodities to specialties defines the need for relationship management.

⇨ The concept of fixed prices for fixed products in arms-length transactions does not convert to competition based on enriching the customer in a relationship over time without profoundly changing the way we do business.

⇨ Balancing relationships across the three principle dimensions can be helpful to interprise leaders in positioning themselves in today's highly competitive world.

THINGS TO THINK ABOUT

☐ How does the change from selling commodities to specialties define the need for relationship management in your business?

☐ Do you pay fair, variable rewards to your suppliers for the enriching service and knowledge which they give you?

☐ Are your company's efforts in information systems, training, and hiring tailored to support the interactive work you develop with your customers and suppliers?

CHAPTER 12

ENRICHING CUSTOMERS

T he enrichment dimension of the interprise relationship model derives from the concept of providing not only a product or service, but total enrichment to the customer.

This is often understood to be the equivalent of providing customer value. On the surface this is a fair substitute of words, but it can lead to false inferences. For instance, the concept of providing customer value and "value adding" activities in the enterprise is often also equated. A "value adding" activity carried out by the enterprise usually refers to costs added in the generation of goods and services that may not always enhance the finished product. This is significantly different from customer enrichment, which can only be measured in the context of the customer. The same product or service can provide different amounts of enrichment to the customer depending on the circumstances in which the customer uses the product. For instance, a plane seat increases in value to the customer the nearer he or she is to the date of departure. Airlines sell their tickets accordingly, so the highest prices are paid for same-day flights.

Enrichment is situation dependent; it is not under the supplier's control, and sometimes it is not under the customer's control either. However, it depends on more than luck. Enrichment is likely to be

121

based on experience over time rather than only at the instant of the transaction. It will probably require a combination of factors, including information and service, and may or may not include a physical product.

Earlier, we showed how the supplier must interactively connect with the customer to enable customer solutions. Now, we examine the various forms that enrichment can take. We have divided each dimension of the interprise relationship model into three parts. They represent a progression of activities in the enrichment dimension:

- Valued but non-measurable customer enrichment
- Measurable customer enrichment
- Enrichment of the customer's customer

In this chapter we shall discuss the first two thirds of the enrichment axis. The division into three parts is convenient for our purposes here, but it is not rigid. Business situations that fall between two segments are quite possible.

NON-MEASURABLE ENRICHMENT

Figure 12.1 The Enrichment Dimension:
Non-Measurable Enrichment.

The first third of the enrichment dimension involves things that suppliers do which customers value, but from which the supplier derives no increased revenue. The enrichment is not measurable in quantifiable ways, but there is a perception of value that the interprise can use to capture and retain customers. For example, most consumers value getting a hotel suite for the price of a regular room. They enjoy the comforts and feeling of luxury the suite affords, but they would not want to pay for it. The experience, however, may well induce a customer to return to the hotel, in the hopes of repeating a pleasurable event.

The idea of *things the customer values, but does not want to pay for* seems at first anomalous, yet it is an important part of the enrichment dimension. Providing amenities, or service, or information without charge, is the basis for building many customer relationships and, more importantly, customer loyalty.

> Frequent flyer miles were designed to have no tangible monetary value to the receiver, but they are good for tax purposes and attract the consumer back again and again to the airline which offers them. The concept has worked so well that the programs have been copied by other types of companies, among them, American Express miles, Diners Club points, AT&T true value points, car rental, and many hotel programs.

These relationship-building activities are the simplest kind of enrichment. Some can be measured, like the conversion of frequent flyer miles into a free ticket. The real value, however, is more elusive. When you cash in a mileage award, you have limitations on seat availability, fewer rights and privileges, and are in fact occupying what the airlines have statistically anticipated will be a vacant seat. These intangible enrichment benefits are defined as non-measurable not because their value cannot be calculated—it can—but because their main function is to engender good feeling. They represent the warm and fuzzy part of the enrichment relationship.

Non-measurable enrichment is the most elementary relationship building step you can take. You provide customers with something they like, and that brings them back. This increases your revenue and attracts customers, but does not have an impact on their bottom line goals. A piece of jewelry engraved with a name or message is more special to the customer, but not more valuable in dollar terms.

The publisher Van Nostrand Reinhold offered our previous book to bulk customers at a price schedule that was not significantly different for mini-customizations or vanilla books. Westinghouse and Dow Chemical, among others, chose customized editions. Each special edition had a slightly different cover, and a page or two with the company's message. The publisher found that while customers did not want to pay more per book for customized copies, they would order more books if the books were customized. Companies liked the opportunity to put in a special message, and ordered enough to cover the additional costs. Indeed, we watched several organizations change their order to three and four times the number of books discussed initially because of the customization.

Can you increase revenues in your business without raising prices by offering customers customized opportunities that do not add cost to your operations? There are many forms of simple enrichment, and not every idea is applicable to every business. This simple form of customer enrichment will not make a difference in all businesses.

The use of special programs such as caller ID or Direct Inward Dialing enable a company to know who is calling them, to address the customer by name when he or she calls, and to use the computer data to automatically display a history of prior individual customer interactions. Some firms will connect you automatically to the same person you spoke with the last time you called. That person sees your name, call record, and customer experience on a computer screen before they pick up the phone and greet you by name.

The concierge service of Platinum American Express keeps a file of a customer's preferences. They can rapidly find the type of gift a customer likes to send, and will retain a record of gifts sent to different recipients.

An important aspect of this first third of the enrichment dimension is the perception of trust and reliability. These are things which customers value in a relationship, and they respond to it with loyalty, if not additional money. The supplier benefits from the loyalty engendered in the customer, and the customer benefits from the reliability of the relationship.

> Nissan sells the high-priced Infiniti luxury car with an extraordinary service package. They provide a sixty-month, five-year bumper-to-bumper warranty; they will supply a loaner car no matter how long the repair is expected to take; they will provide transportation when a customer has an accident or simply breaks down. Nissan, Lexus, and a select few others, have moved beyond selling cars; they are selling transportation when and where the customer needs it.

> Virgin Airlines is establishing a reputation for anticipating and satisfying customer needs in a high class way. For a trip to London, they offer first class service for business class prices, including a limousine pick-up from your home or office. As they enter other business domains, they will carry the customer loyalty and expectations built through their airline service.

Customers choose to do business with companies they can trust; trust is something the customer values, but does not put a price on.

MEASURABLE ENRICHMENT

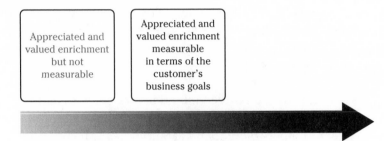

Figure 12.2 The Enrichment Dimension:
Measurable Enrichment.

The second type of enrichment can be measured by the customer in terms of their bottom line goals. By this we mean not only that the enrichment can be calculated—if only approximately to the first digit and a number of zeroes—but that it makes a significant difference in the bottom-line metrics that a customer uses to manage and run their business. Most of the time this will mean bottom line profit, and it is up to the customer to choose which metric to use. For instance, some not-for-profit customers may be cost driven and able to measure the financial impact of a supplier's enrichment; others may value the supplier's ability to help them reach more customers, or spread their message more broadly.

For profit-driven customers, there are two general ways to provide enrichment that they can measure and will value:

• The supplier can reduce their customer's business costs—enrichment through savings.

• The supplier can increase the customer's benefits from its business—enrichment through profits.

Ways to decrease the customers cost of doing business include:

• Reducing the amount of inventory necessary, perhaps by using a manufacturing resource planning system (MRP).

- Reducing the cost of supplies, perhaps by using a purchasing system or negotiating relationship contracts with suppliers and eliminating waste between the two organizations.
- Showing how the same amount of business can be done with fewer people, perhaps by introducing a schedule management system.

There are, of course, many ways to help your customer cut or contain their costs, and many businesses are founded solely on this concept. It is crucial that the supplier is able to measure the savings that result from their efforts. It must know the customer and its operations in great depth in order to both enrich the customer and learn how the customer measures the value of that enrichment. Typically, enrichment is not instantaneous and must be measured over time. Even after initial success, the supplier should continue to look at goods, services, and interactions so that it can adapt as the customer's needs and situations change. To ensure that the customer attributes measurable benefit to the efforts of the supplier, requires that the supplier evaluate the financial benefit from the customer's perspective in the customer's context.

The second approach is to increase the customer's benefits of being in business. Here the supplier should focus on two distinct categories of opportunity:

- Increasing the penetration of the customer into existing markets.
- Opening new markets for the customer.

We examine the first idea here and the second in Chapter 13, where we discuss enriching the customer's customer. Neither concept is exclusively limited to the middle third of the enrichment dimension; both are achievable in either the middle or last third of that dimension.

To assist a customer in achieving greater market share, the supplier seeks ways to make the customer more attractive to the market. This can be achieved using various techniques:

a) Reduce the time it takes to bring your customer's goods and services to the market.

First, the supplier must check the assumption that by getting to the marketplace earlier, the customer can achieve a bigger share. The supplier must check that the customer's product or service has a market, and that more rapid delivery of the product or service would be valued. The supplier is not providing enrichment if the customer's customer does not want the product or service, or does not value the extra speed. When a supplier connects with its customer, they both need to learn not only how the supplier can help, but also the potential impact of that aid. If the customer fails to achieve the enrichment they sought, the supplier may suffer, even though it contributed its best efforts. This is one of the factors that leads world-class suppliers to seek out world-class customers. A supplier who develops good enriching capabilities proactively looks for customers who are capable of using and valuing those capabilities.

b) Help the customer increase the quality of their goods and services in order to capture more market share.

The assumption here is that the customer is still in a market where quality is not yet a universal standard and can win business. Unique levels of quality and service are still possible in most industries, as Lexus and Infiniti demonstrated when they set new standards for luxury car service and quality.

c) Help your customer integrate product, information, and service to better serve their clients and capture more of the market as discussed in Chapter 6.

In order for this to be successful, your customer's customer must value integration and relationship building. Mass production, product-focused commodity businesses are, by definition, not headed this way. On the other hand, when a supplier can help a customer turn a commodity business into a specialty business, the supplier generally increases its revenue. The United Kingdom's Virgin Airlines and Brazil's TAM both specialize in attracting customers who value an integrated specialty product that is easily distinguished from the no-frills commodity versions against which they compete. Levi Strauss's personalized jeans are directed at a similar market.

d) **Help the customer customize product-service combinations to meet unique customer niche needs as discussed in Chapter 7.**

This is different from the above, since in c), the customer wants integrated goods, information, and service. Here, the customer seeks an individualized product or service, and the supplier helps the customer in cost-effectively producing customized, or one-of-a kind, goods, and services. In essence, the supplier helps the customer achieve economies of scope, rather than scale, to customize goods and services without significant additional cost. To do so in a way that the marketplace will value is the critical test of success.

Part 1 of the book showed the different approaches to providing value to customers, whether by using products and services as platforms for additional value, by customizing products and services, or by market fragmentation. In this chapter we have begun to explore a systematic approach to providing that enrichment. The trailblazing leaders found their way by intuition, perseverance, and simple trial and error. From their efforts, an overall approach is now emerging.

SUMMARY POINTS

⇨ Customer enrichment is measured in the customer's terms, not yours.

⇨ There is no absolute sense of enrichment. The same goods and services can and will have different enrichment values based on the customer's experience and situation.

⇨ The enrichment dimension identifies intangible enrichment which can be used to attract and retain customers.

⇨ The second third of the enrichment dimension is about getting connected with customers by understanding their real needs and what will make a difference for them.

⇨ Measurable enrichment is described in the last two thirds of the enrichment dimension.

⇨ The enrichment value a customer measures can come from cost savings, increased market value, or the opening up of new markets.

⇨ Many of the enrichment opportunities come from creating specialty items, or customizing solutions for the customer that
often include combining goods, information, and services in
unique ways to provide enrichment value over time to the customer.

THINGS TO THINK ABOUT

☐ Can you list enrichment possibilities in the first or second
thirds of the axis?

☐ Are there enrichments you would want your suppliers to bring
you?

☐ How would you prioritize opportunities with customers as you
develop enriching methods for them?

☐ How would you manage customer interactions for enrichment?

☐ How would you prioritize your opportunities with suppliers as
you develop enriching methods for them?

☐ How would you manage supplier interactions for enrichment?

☐ What restructuring would you anticipate in your company in order to deal with these efforts?

CHAPTER 13

YOUR CUSTOMER'S CUSTOMER

In the final third of the enrichment dimension shown in Figure 13-1, we focus on how to provide measurable enrichment to the customer's customer.

Appreciated and valued enrichment but not measurable

Appreciated and valued enrichment measurable in terms of the customers bottom line goals

Appreciated and valued enrichment measurable by the customer's customer

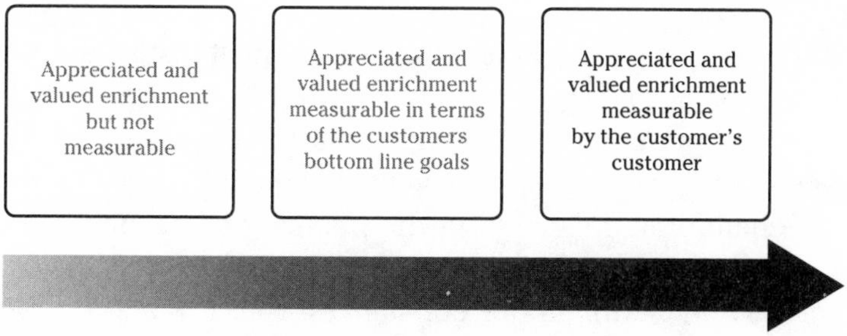

Figure 13.1 The Enrichment Dimension: Enrichment Measurable by the Customer's Customer.

The old competitive world was managed by the "push" principle, with goods and service decisions pushed up the supply chain by suppliers. Even if the supplier of product or service consulted with customers, it was the supplier, not the customer, who decided on

131

the design and price of the goods or services provided. In the financial accounting methods of that push-driven, value-adding chain, when the supplier added cost, it was assumed that value would follow. Competition has moved from the push-driven value-adding chain to the pull-driven value-adding chain. Now, the customer makes the decisions regarding the functionality, configuration, and the price he or she is willing to pay for the supplied product or service.

> When Unisys pioneered its "customerize" program, they set out to provide significant enrichment to their customers. They decided that instead of selling information and computer systems, they would sell their ability to help customers become more effective with their customers. To this end, Unisys helped airlines manage complicated reservation systems and individualized frequent flyer programs; they assisted banks with depositor services; they aided welfare and other government agencies with their constituents, and helped phone companies increase the number of phones and service per phone.

Helping your customer help its customer to see bottom line benefit, is the first step in visualizing an enrichment chain. The phrase "value-adding" is generally used when work is performed and cost is generated, and is seen from a supplier and cost perspective. Enrichment-adding activities, however, view the processes of the customer's chain of suppliers from the perspective of the value they provide to the customer. An enrichment chain starts with a knowledge of the end-customers or consumers, and how actions at each link in the chain will provide them with measurable enrichment.

> Johnson Controls was mentioned in chapter 6 as an example of a company which has used their product as a platform to provide service. In doing so, they advanced from working with customers—the contractors who built buildings—to the contractors' customers, who own and operate the buildings.

LIVING YOUR CUSTOMER'S EXPERIENCE

As discussed earlier, business success today requires coordination of the interfaces with both customers and suppliers, and within the company. To do so requires an understanding of how your customers and suppliers succeed in their business, as well as establishing a dynamic and flexible operational structure within your own company.

> The recently formed Cambridge Technology Partners (CTP), under the leadership of CEO James K. Sims, is thriving in the information systems business not just by selling information systems, but by selling customer outcomes, and even customer's customer outcomes. Sims notes that only 11 percent of business process reengineering projects succeed. He can also quote surveys which conclude that of the $80 billion of information technology systems projects in 1995, 31 percent were canceled, and only 16 percent of the rest were successful. As a result, CTP refuses to bid on a project where the customer sends a specification, because they know that by the time the work is done, the specification will be outdated and the effort will have been wasted. They work with the customer's people to thoroughly understand the customer's business so that a well-functioning prototype information system can be rapidly built, then turned into a flexible production system.

> Sims attributes CTP's phenomenal success to people and people skills. He suggests that his company's successful customer enrichment is dependent on knowledge in three areas:

> 1) Information systems
> 2) Behavioral dynamics
> 3) Specific industry and organizational knowledge

> CTP claims expertise in 1) and 2) above and assumes its customers bring in 3). Thus, they are able to function in a

wide variety of industries and industry sectors, and are rapidly expanding their global reach. Sims suggests that their knowledge in 1) must be renewed frequently. They continuously reinvent the information systems techniques they need because technology changes so fast. The mainstay of their expertise, however, is in 2), which deals with relationship building and facilitating. They interview ten people for every one they hire. They find that most applicants are technically qualified, but not well-versed in the people skills Sims considers essential in working with CTP's customers. The company will spend about 70 percent of their interview time on people and behavioral skills, seeking individuals who can communicate, work in teams, lead and follow, and who can learn to facilitate the success of others. Because they believe in the people who work for them and recognize that they need them to drive the business, CTP is not managed as a monolithic, hierarchic organization, but has operated over twenty-one business units in seven countries even when there were only 1,200 people in the whole organization.

To better understand their customers, CTP uses a few weeks of joint seminar discussions to help the customer uncover what information systems will make a significant bottom line impact not only for the customer, but for the customer's customers. CTP is in the enrichment business. Both its knowledge of information systems and the formidable people skills of all its employees are crucial in enabling the company to partner with customers in order to define meaningful and real enrichment chains.

ENRICHING RELATIONSHIPS

The essence of enrichment in the third part of the enrichment dimension is to introduce products and services which closely match the spectrum of needs for various customers, and to help your customer understand its customers and enrich them in new and potentially unique ways. The customer seminar that CTP uses is just one of many ways of getting to know your customer and its customer. Another method is to station employees permanently at the customer site so they become a part of the customer's product and service development team. Organizing meetings and workshops, or communicating extensively through electronic media also works. The type of method doesn't matter; it is the creation of a relationship over time that will give you the knowledge to help your customer better serve their customer.

In this interactive, two-way relationship, it is important to learn how your customer perceives your organization—whether you are perceived as a facilitator or an impediment, or perhaps a mixture of both. You should strive to maintain a relationship close enough to allow honest and frank exchange of views. This enables you to understand where the opportunities in the relationship are and could be.

In developing this close relationship, a problem that some suppliers have come up against is that customers might not have or want to acquire the requisite skills to benefit from what the supplier can provide them.

> The electronic information and data systems company, EDS, routinely offers to form a partnership with its customers in order to increase their ability to provide strategic use of information as a competitive weapon. Their customers frequently discover, however, that they do not have the technological skills to fully exploit the systems EDS can offer. So, when EDS makes this type of strategic arrangement, it will often hire all of the customer's information systems people, then assume responsibility for providing information where and when it is needed in a virtual organization relationship. Some would call this a strategic alliance or outsourcing, but these names are usually applied to arm's length relationships. EDS's people act and feel like customer employees. They know they work for EDS, but they are assigned to think and act like the customer's employee, working to benefit the customer. When EDS takes on the customer's point of view, it

helps the customer focus on its own enrichment chain. In this way, they have built successful relationships with many companies, including Bethlehem Steel. EDS acts as a member of the Bethlehem Steel senior team, helping them focus on how they can enrich their customers.

EXPANDING RELATIONSHIPS

The control systems company Landis and Gear thought it was in the plumbing and air conditioning controls business, selling to building system contractors, until they discovered that their contractor's customers really wanted them to provide climate control and energy efficiency for their buildings. Landis and Gear not only rethought their products and processes, but also who they considered their customers to be. Now they take contracts to operate buildings and shopping centers for more profit than they previously made selling the control systems used in those buildings.

The key to this type of enrichment lies in understanding what customer chains you are part of, what each customer values, and how to reengineer your business to be able to provide that value reliably. Frequently, this is worth far more to the customer than was obtainable from the commodity systems and spare parts bought by purchase agents. Furthermore, it can lead to new and unexpected business opportunities.

The Japanese Mayakawa organization discussed in chapter 4 works with their food industry customers to produce cooling equipment. Their engineers learn about the processes that food undergoes throughout the entire value-adding chain, thereby enabling them to take advantage of business opportunities derived from understanding the needs of their customer's customers.

Magazine publishers collect information about their readers to better serve them and meet their interests.

They also use the information to gather advertising revenue. Today, they routinely sell access to their database to marketing companies for very lucrative fees; they are selling access to their understanding of their readers.

These companies have gone beyond a relationship with their customers, to interacting with their customer's customers. In doing so, they have expanded their range of opportunity. The principle is to examine your business competencies and ask if your customer's customer might gain significant enrichment from direct access to your expertise or information.

Understanding the circumstances of the customer and its customer and determining what it is trying to achieve, suggests selling outcomes, not products and processes. If you worked for a machine tool manufacturer, would you have thought of selling management of factories, instead of the tools which constitute the factory?

Cincinnati Milacron discovered that what many of their customers really wanted were efficiently operating factories, not just the best equipment. They found that they make more revenue and are best equipped to help the customer better serve its customer by being a part of its business instead of a supplier to it.

The concept of selling outcomes by joining with the customer to help enrich the customer's customer is a powerful and effective approach to expand business and profit from the new competitive environment.

SUMMARY POINTS

⇨ To provide measurable enrichment, businesses must form meaningful relationships with their customers so they can help the customer better serve its customers.

⇨ Joint seminars and workshops, stationing employees at the customer's site, and electronic communication are some ways a supplier can learn its customer's business.

⇨ Companies should look at the things they sell to determine if they should be selling managed outcomes rather than products and services.

⇨ Understanding how it impacts the enrichment chain will help a company take action to make the chain more effective.

THINGS TO THINK ABOUT

☐ Do you know how your business customers make money with their customers?

☐ Do you know your individual consumer or end-user lifestyles well enough to know how you may be a part of them?

☐ If the answer to any of the above is no, how could you find that information?

☐ If you found out, how would you know the information is reliable? What opportunities could there be for you with your customer's customers?

☐ Would your people know how to become effective parts of your customer's customer satisfaction processes?

CHAPTER 14

PRICE FOLLOWS VALUE

Relationships are the key enabling factors in building an interprise. Without appropriate reward, however, relationships will not last. As companies move from the traditional arm's length supplier roles to providing added value in new ways, the methods by which the supplier is paid will change. In this chapter, we examine the way in which customers and suppliers can share both risk and reward in the new environment.

We have suggested that movement in any one dimension of the interprise relationship model will necessarily be connected to movement in the other dimensions.

Movement up the value-adding axis, for instance, generates corresponding movement up the reward axis. When the customer takes a risk to gain enrichment, the price paid by the customer follows and is linked to the risk and reward, which are shared with the supplier. Since price must follow value, it will depend on the context for each case and its circumstances. The reward dimension introduces a way of separating and measuring the degree of sharing of, what some call, a common destiny.

As described in chapter 8, Goodyear supplies tires to air-planes by taking care of all the tire service for the airline. They charge not per tire, but per landing.

R.R. Donnelley specializes in rapidly putting together and publishing financial prospectuses for various types of business offerings. For the customers, time is very valu-able to the process. Therefore, Donnelley does not charge according to the cost of the printing, but takes a fraction of the value in time saved for the customer.

A SPECIAL RELATIONSHIP

Figure 14.1 The Reward Dimension: Sharing a Special Relationship.

As illustrated in Figure 14-1, the first third of the reward dimension deals with special relationships, where there is sharing between customer and supplier, but no significant change in the flow of money between organizations. In this type of relationship, it is more a matter of doing things which improve the other's operating posi-tion. When two companies work together to develop product spec-ifications for the products to be supplied from one company to another they can frequently reduce time and costs while increasing quality. Chrysler, for example, routinely invites suppliers in to work with them on design.

Goodyear has developed a special version of its tire sim-
ulator to interface directly with its customers. At the site
of a customer who operates a fleet of vehicles, the sales-
person enters the characteristics of the fleet and its use
into a computer, and receives immediate data on the eco-
nomics and other information for different Goodyear
tires.

This is a stretching of the one time arm's length transactions into
a close relationship over time. The supplier and customer work together to find a mutually beneficial way of operating. This also
occurs when suppliers agree to be just-in-time suppliers (JIT), or as-
sume responsibility for entire subassemblies instead of components. There are similar increases in customer enrichment when
suppliers agree to provide a variety of products on a just-in-time basis, with prices based on large volumes, while allowing significant
variation in the specifics of orders.

Harley Davidson has a number of its suppliers in York,
Pennsylvania, working as JIT suppliers for a variety of
products. Harley routinely adds or cancels part orders
with very little notice. The suppliers are happy because
they have a relationship that is broader than a single
part, while Harley is happy to have the JIT deliverables
and a relationship which is more agile in terms of long-
term commitments to individual parts. The interactive relationship with shared information makes feasible and
worthwhile what would otherwise be unacceptable behavior in the one-transaction-at-a-time, arm's length marketplace of mass production. Teams of Harley engineers
routinely meet with supplier engineers and executives to
discuss plans and opportunities.

Harley has also asked suppliers to work together as teams, because they find it easier to interact with a team than with each supplier individually. Similarly, many automotive companies are
starting to require that their suppliers work with other suppliers to
provide a subsystem, such as a complete auto interior or power
transmission system. This manifests as a major structural change

in industry towards the formation of supplier webs that are actively linked together. Such webs can save time and money and can increase quality by operating in concurrent modes, sharing information with each other and the customer.

THE ENRICHMENT PRICING SEGMENT

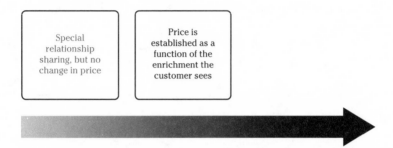

Figure 14.2 The Reward Dimension: Price Becomes a Part of the Customer's Enrichment.

In the second third of the reward dimension (see Figure 14.2), in which price follows value, deeper relationships are formed that take advantage of interactivity in a way that goes beyond arm's length transactions.

> For instance, an engineering firm suggested to its customer, an automotive firm, that they could help make cars stronger, safer, and more economical. They jointly calculated a potential benefit of $350 million in the first year of the new engineering system. At first, the supplier offered to install the new technology and get it into production for $100 million in an arm's length transaction. The customer responded with a price-follows-value deal. They asked the supplier to install the system without a fee in exchange for 40 percent of their savings in the first year, and a declining percentage thereafter. (This story is true, but the numbers are fictitious.)

Interprises that offer solutions to their clients frequently complain about the difficulty of selling a solution for a high price. In part, this resistance comes from a lack of customer experience with trustworthy claims made by suppliers. If a supplier really believes it can provide a valuable and enriching solution to its customer, it may actually receive a higher return and an easier sale by agreeing to a performance-based price. Price then also becomes a function of the enrichment value the supplier provides to the customer. Small, entrepreneurial firms who cannot afford to finance an investment while waiting for the return will experience the down side of this approach. They need enlightened, risk-taking, banking relationships. The Japanese Keiretsu is a permanent group of cooperating companies, who often hold equity positions in each other's companies. According to some, the main reason it is customary for a Keiretsu to have banks in their cooperative is to meet the need for just this kind of enlightened, financial risk-taking.

Wider adoption of this approach is also hindered by the fact that some management have difficulty eating their own cooking. If unreliability in a forecast would result in not being paid, executives—who clearly do not even believe their own marketing people—will not make a pay-for-value agreement. This type of leader will find it hard to make the transition to relationship-based competition. If your company does not believe in and is not willing to stake its future on its own marketing claims, it will not convince others. Customers in the interactive business world now select suppliers based on relationship values. If businesses don't or won't play in this game as an organization that keeps its word and stands behind its marketing claims, they will do no better than survive in a commodity-only, low value-added, product-focused marketplace. They will not succeed in the enriching, relationship-based, price-follows-value marketplace.

A potential problem for price-follows-value arrangements is that the customer may turn its back on a potentially mutually beneficial gain because it is afraid of such an innovative approach. Such businesses will continue to insist on fixed price payments, or the supplier can ask the customer to reimburse the fixed costs of improvements in return for a more modest share of the increased profit.

TRUST AND SHARING

Companies are currently exploring ways of measuring in an audit-worthy way their contribution to their customer's revenues. We feel sure that the entrepreneurial spirit that is leading the way in this current environment will soon find acceptable approaches to evaluating customer enrichment. Solutions will certainly include the following two components:

- There will be more open exchange of information, including confidential business and financial information.
- The existence of a long-term relationship between companies will enhance the chance of finding mutually acceptable formulae for measuring enrichment. If the supplier wants to do business with a customer in the future, and vice versa, both are powerfully motivated to set up trustworthy arrangements.

Unfortunately, data on specific reward-sharing methods is still sparse, since this information is often confidential. A 1992 annual survey of CEOs in the electronics industry, produced jointly by the consulting firm Ernst and Young and *Electronic Business* magazine, reported that two-thirds of the CEOs surveyed say that alliances "figure prominently in their strategic plans." Approximately 81 percent believed that marketing alliances would increase over the next five years, while 77 percent predicted access to technology and joint product development work.

Businesses have to examine the solution sets they are providing to find out if there is a clear and suitable method of determining and accounting for increased customer enrichment. If this is the case, then price-follows-value arrangements can be a lucrative option to pursue. Trust is the key enabler in these relationships, and agreements defining the ethical and financial expectations of what will take place and who is responsible will contribute significantly to a successful venture.

THE SHARED RISK AND REWARD SEGMENT

Special relationship sharing, but no change in price	Price is established as a function of the enrichment the customer sees	Price is replaced by a sharing arrangement based on the value provided to the enrichment chain

*Figure 14.3 The Reward Dimension:
Shared Risk and Reward.*

The final third of the reward dimension (see Figure 14.3), represents what might have once been called a joint venture. A permanent joint venture, where all the terms are spelled out from the beginning, is less likely to be an adequate mechanism for the fast-paced, quickly changing, competitive world we live in today. Opportunities are growing in number and shortening in duration, thereby rendering joint ventures, which are permanent and expensive to form, more and more inadequate. That precise legal arrangement can now often be replaced by a virtual organization relationship.

As discussed in chapter 16, virtual organizations are time-based and opportunistic; they last as long as the opportunity. For instance, some organizations have created a model legal contract that can easily be deployed by filling in the blanks when the need arises. They find that the more agile and easily formed virtual organization structure matches the needs of a changing marketplace and opportunity set, Rather than relying on sharp legal practices based on an adversarial system, success requires trust, and lawyers that have the motivation and business acumen to find new ways to make successful deals. The aim is to succeed together, rather than to score the best in a one-time deal.

In this way, when two or more organizations have a relationship which allows them to work together, they may see an opportunity in which they agree to share costs and contributions and rewards and profits. Wide varieties of latitude exist in the types of splitting arrangements which can be used, but a high order of trust, shared ethical expectations, and values are required for success.

> For instance, when DuPont and Lockheed agreed that they each had too many business opportunities to handle alone, they had their lawyers create a model for virtual organization formation when the occasions arose. They then began a series of meetings in which they described both capabilities and opportunities they thought might develop.

Sharing arrangements in virtual organizations vary from formal systems to handshakes. In today's fast paced and agile world, things do not stay the same for long, and experience shows that new opportunities often arise out of virtual organization combinations. Organizations that have decided to cooperate for a limited opportunity often discover that together they can launch projects in new business areas.

> Hughes Electronics, in seeking ways to bring its new eighteen-inch satellite dish to market, has launched a number of equity-sharing relationships, most successfully with the direct satellite systems company DSS. Because of this initial relationship, DSS is now a partner in additional ventures with AT&T, Microsoft, and a variety of entertainment and sports networks.

The business world is moving to shorter time horizons for deals with longer term relationship implications. The winners will protect their partners, not take advantage of them, and will recognize the time-based and opportunistic nature of reward sharing arrangements from the start.

SUMMARY POINTS

⇨ Price alone is just part of the relationship; customers will value innovative arrangements in delivery, coordination, and innovation to provide them with enriching solutions.

⇨ Trust and the development of shared ethical understanding are key components of success with the reward dimension in which price follows value.

⇨ When you can determine the enrichment you have provided to a customer, you may want to charge a fraction of the enrichment as the price.

⇨ Price-follows-value pricing can be experimented with. Many companies start with cost reimbursement and partial enrichment based payments.

⇨ Equity sharing is a realistic possibility and should be considered.

⇨ Seek out world-class partners who share your ethical expectations and recognize the value of long-term relationships even with short-term opportunities.

THINGS TO THINK ABOUT

☐ Do you know what added value your product or service brings to your customer?

☐ Are there efforts you could make which add value to your customer, and which are low expense items for you?

☐ Are there opportunities to turn your first third value-adding efforts to middle third efforts, thus bringing you a better return?

☐ Do you know customers well enough to perceive last third reward opportunities?

☐ Are you actively working with chosen suppliers to see if they can bring you significant added value?

☐ If so, would you know how to reward them?

CHAPTER 15

BUILDING
LINKAGES

I n building relationships, businesses aim to better understand each other, and to learn how to make win-win relationships with two or more players. The linkage dimension describes the way people and systems communicate with each other in order to achieve mutually supporting, but often distinct, business objectives. Implementing systems and work processes in the linkage dimension goes hand in hand with progress in the enrichment and reward dimensions. The linkage dimension, however, should not be mistaken for technological linkage and integration. Electronic interaction is a means, not a goal; the goal is to enable close, effective relationships between people and their organizations that open new opportunities for mutual profit.

New technical capabilities are pushing organizations to link their processes more and more intimately across a spectrum of industries. From retail stores like Wal-Mart, K-Mart, and Sears, to electronic companies like Gateway, Dell, and Motorola, to the automobile industry, the daily linkage of work processes has become routine. Because of this linkage, companies are finding that their relationships develop from doing the old thing faster to doing entirely new things. As relationships move up the three segments of the linkage axis, they correspondingly move up the enrichment and reward axes.

149

VALUED LINKAGE

Figure 15.1 The Linkage Dimension:
Valued but Not Measurable Linkage.

As with the other two dimensions, the linkage dimension is divided into three parts. The first third deals with linkages that occur naturally, as businesses talk to the customer or supplier, sharing data and giving feedback (see Figure 15.1). Linkage on the first third may consist of communicating through the Internet, or at a meeting at a trade show. Results of this type of activity are valued, but not measured. They feel better than the old arm's length way, but people have trouble articulating why. Intense interactive communication does not usually occur in this third; nevertheless, both parties may learn more about each other than in isolated transactions through the use of online information services, business brokers and representatives, and salespeople.

> A few years ago, Seven–Eleven in Japan began asking their store clerks to enter simple data about their customers after each purchase. They entered age and sex of the customer, and the items purchased. This data base of customer buying patterns was then used to select the products that were offered, and create the layout of the stores.

While this approach is an improvement, it is not as effective as interacting with a customer. Interaction allows a business to learn how a supplier's efforts can be helpful and share with the customer information on the supplier's capabilities and constraints so that they can mutually modify their relationship to improve performance and eliminate waste for both. This latter approach is commonly used by companies working in the second third of the linkage dimension. IBM uses it in their division which makes PCMC cards, as does Unisys in their customerizing program. Many companies will begin to foster linkage by holding supplier meetings. In conducting such meetings, businesses should have a clear idea of whether they are aiming for linkage in the first or middle third of the dimension.

MEASURABLE AND REWARDED LINKAGE

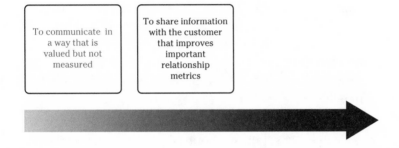

Figure 15.2 The Linkage Dimension: Measurable Linkage.

The middle third of the linkage dimension (see Figure 15.2), is based on seeing a measurable increase in effectiveness of the relationship. It is becoming common for organizations to hold meetings and exchange data by regular, electronic, or other means to dramatically reduce time, lower costs, increase quality, and generally improve the collaboration in a key metric of performance. The improvement is made by the people in both organizations, but is enabled by exchanges of information, both in personal meetings and by electronic means.

A common method of middle third linkage is an electronically coordinated logistic system. Such systems have been implemented by many companies; for example, Motorola's Schedule Sharing System, and Wal-Mart's well-publicized electronic linkage with its sup-

pliers, both of which are described in chapter 4. This linkage creates measurable benefits for the customer. The supplier's motivation to find new cost savings is often increased when the customer's
increased profit, or reduced expense, is shared with the supplier.

Middle third linkages can occur not only between companies, but
also between a supplier and an individual consumer. When a company produces a new product or service, first third linkage thinking
suggests an advertising campaign aimed at showing existing and
potential customers why they need the new product. Middle third
linkage thinking will lead to free trials or samples, and follow-up interviews.

> The Saturn has had an impact on customers in part be-
> cause the corporation decided to build a middle third re-
> lationship with customers wherever possible. Saturn
> factory workers routinely call new owners and say, "I
> made your car. How do you like it?" They listen to what
> the customers say, and follow up on complaints, giving
> customers the sense that they have joined a caring
> group.

> Lee Iacocca has talked about how Chrysler recovered
> from a bad situation when it had difficulties with its SJ6
> engine. They called 1.4 million customers and began the
> conversation with a genuine expression of how sorry
> they were for the problem they had caused. They then
> went on to say how they would fix it. The customers re-
> acted positively to the personal call and genuine expres-
> sion of sorrow by the company.

Because of the unique consumer perception of their product,
Harley Davidson customers start with an established sense of iden-
tity with the company and with other customers. They expect to be
in a relationship with the company and with each other. SAAB owners have similar feelings as well.

When an unhappy customer begins a conversation with a company, the opportunity exists either to build a linkage relationship

between customer and company or to destroy an arm's length one. Businesses that consider customer complaints as a chance to build positive, long-term relationships, will prosper, and those who do not will be in trouble, since customers are beginning to realize how they can benefit from middle third linkage relationships, and will soon expect this kind of treatment.

VIRTUAL TEAMS

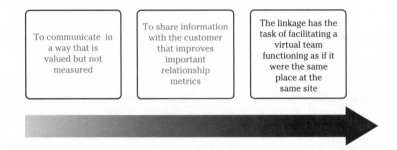

Figure 15.3 The Linkage Dimension: Virtual Teams.

In the third part of the linkage dimension, we look at how information and people intermingle to form an integrated virtual team across organizational lines (see Figure 15.3). In this third, people in separate businesses work together as members of the same team, using interpersonal communications both face to face and electronically to facilitate and increase the effectiveness of the teams. Groupware information products such as Lotus Notes, electronic data interchange (EDI), and the Internet form the basis for electronically-aided collaboration. The main benefit is in the increased effectiveness of the people in a virtual organization, working together in different geographic locations, across multiple disciplines, and with the same or better effectiveness than if they were in the same location at the same time.

R. R. Donnelley's financial services printing group has established an expertise in helping geographically dis-

persed groups of people in different companies work together to create the financial documents for events such
as public offerings or debt issues. Donnelley claims that
they can rapidly and effectively integrate the people to
produce the required documents in a short period of
time.

Although there is much fanfare over the technology of interactivity, there are only a few interprises that routinely achieve top third
linkage. The most notable examples are in movies, news media and
financial services. In these businesses, virtual teams are frequently
formed for projects, or to cover particular stories that span the
globe, or to put the finances together across multiple markets for
brokerage firms and international funds.

The wave of big company mergers and alliances in the mid-1990s
reveals a widespread perception that a substantial payoff lies in the
third part of the linkage dimension. There will be opportunity for
more effective utilization of this type of linkage as the new global alliances in travel, telecommunications, and other services make
their presence felt. Businesses are only just beginning to create the
truly global virtual organization that the last third of the linkage dimension foresees.

SUMMARY POINTS

⇨ The purpose of linkage is to enhance the effectiveness of people
in meeting business objectives.

⇨ It is important to increase the effectiveness of relationships by
sharing information using both meetings and information exchange.

⇨ The real benefits of linkage that integrate disparate teams of
people in global virtual organizations are just emerging.

THINGS TO THINK ABOUT

☐ Are you mostly absorbed in the technology of electronic linkage within your business and with customers and suppliers, or in the uses to which you put that technology?

☐ Have you coordinated between the intensity of intercompany linkage, value added to customers, and methods of being paid?

☐ In your computer network system, does the word "upgrade" mean improve the system while doing the same things for customers and suppliers, or does it mean doing new, better, and different things with them?

CHAPTER 16

VIRTUAL RELATIONSHIPS

In the search for competitive efficiency in the modern, communications-driven, global economy, companies will organize teams that cross business unit or corporate boundaries, in order to reduce costs and to react faster to new opportunities. In most businesses, as much as 90 percent of cost and time derive from the requirements and actions of customers and suppliers. These can only be dealt with by coordination along the whole value-adding chain. Companies must have the ability to accelerate and decelerate if they are to remain competitive in today's business environment. Many companies do not have the time to hire people and build a new core competency when needed, nor do they want the pain of shedding a department or people when they are not needed. The alternative is to move in a coordinated way along the three axes of the interprise relationship model: the supplier adds value to the customer, customer and supplier share risk and reward, and work processes are interactively linked. As businesses move further into the three relationship dimensions, they enter the world of virtual relationships.

A division of a $1.4 billion company that manufactures materials handling equipment, such as forklifts and auto-

mated warehouses, was purchased with a large amount
of debt in a leveraged buyout. The new company immediately went through a cost-cutting and downsizing period
that eliminated their product development capability.
They needed new products, but they did not have the
money to pay suppliers for the design work. Their solution was to convince selected suppliers to work interactively with them to provide the product designs, and pay
on the basis of shared risk and reward. Forming a virtual
organization reduced the potential profit for the division,
and created a new and unusual experience for the managers involved, but they played the cards the LBO investors had dealt them.

WHAT IS A VIRTUAL ORGANIZATION?

A virtual organization is a collection of business units in which people and work processes from the business units interact intensively
in order to perform work which benefits all. It is distinct from a partnership, strategic alliance, or other form of joint venture, where
there may be joint ownership or joint executive control and responsibility, but in which work processes remain isolated and non-
interactive. The virtual organization may be between companies, or
between business units within a company, or both.

Large organizations turn to virtual organization to break down
the functional "silos" created by isolated departments. As IBM went
through its crisis and rebirth in the early 1990s, it put together mul-
tiorganizational and multidiscipline task forces. Small companies
turn to virtual organization to join with others so that they can gain
the benefits of expansion without actually doing so. The unpre-
dictability of military challenges has led the modern military to
make intensive use of special mission-specific task forces, com-
prised of elements of all the functions needed for a particular job.
The military has in place an extensive list of procedures to enable
the organizational flexibility and reconfigurability needed for this.

Virtual organization has existed since humans found that they
needed to cooperate to overcome challenges, but it is only now becoming a widespread mainstream business approach. The Nan-

tucket whalers, who caught and sold a major product in the economy of the mid-nineteenth century, operated as a virtual organization. Not only did each crew member bring his skills and tools, and receive his reward of a prearranged percent of the proceeds, but the captains monitored the movement of whales very carefully, and shared that information only with captains within the Nantucket whaling community. This led to the dominant position of Nantucket in the economy of their time. Then, as now, reputation was paramount. A captain suspected of disclosing the secret data on whale movements outside the community, or a seaman with a reputation for unsupportive or dishonest behavior, would be ostracized and lose his livelihood.

When travel agencies form affiliations with local offices around the world to service their globe-trotting clients, they are using a rudimentary form of virtual organization. They are thinking globally while acting locally. Airlines, such as USAir and British Airways, Delta and Varig, United and Lufthansa, have also formed global alliances to better serve their customers. MCI and British Telecom have joined together to offer their "Concert" communications platform, while Deutsche Telecom, Sprint, and France Telecom, among others, are forming global communications firms. The customer will see only one interface as the global suppliers will integrate their systems to provide unified access. If these relationships were limited to some form of shared equity or coordinated management decisions at the executive level while the workers continued their arm's length existence, the term virtual relationship would not be appropriate. The virtual relationship evolves as the workers in companies team together to coordinate their efforts in common projects.

CHARACTERISTICS OF THE VIRTUAL ORGANIZATION

So far, most virtual relationships are based on the principle that cooperation in all-star teams enhances the competitive capabilities of all participants. The price the participants pay is in learning how to work together, both to share and protect intellectual property, and to respond locally from across the globe with innovation and ingenuity, serving and treating each customer individually. Success means meeting each customer's unique needs, rather than creating a rigid monolith with a hierarchy of rules and procedures.

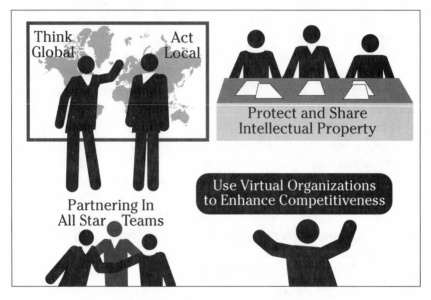

*Figure 16.1 The Modern Virtual Relationship
Linked by Information Technology.*

The modern virtual relationship (see Figure 16.1) is a relatively new organizational model that uses information technology to dynamically link people, assets, and ideas, and is often made up of an opportunistic network of companies that come together quickly to exploit fast-changing opportunities. Since the introduction of the term in our 1991 report to the U.S. Congress to describe a major systemwide innovation, and its subsequent appearance in a 1993 *Business Week* cover story, the characteristics of a virtual organization have expanded. Not every virtual organization has every characteristic, but the definitions below are a good checklist of possibilities and explanations for those who plan to use this mechanism.

CHARACTERISTICS OF THE VIRTUAL ORGANIZATION

- Time-based and opportunistic
- A web of companies who contribute resources
- Greater than the sum of the contributors' capabilities
- Seamless and borderless
- Structured to be dynamic and adaptable
- Virtually vertically integrated
- Totally integrated information and engineering systems
- Linked through interenterprise business and production systems
- Intensive, interactive, and shares information within and among companies
- Aimed at reduced concept-to-cash cycle time
- Aimed at reduced design-to-conceptualization time
- Aimed at one-stop shopping

The most important characteristic of a virtual organization is that it is time-window-based, meaning that it is created to meet a specific opportunity in a defined or estimated time frame. Some virtual organizations may last decades, because they are associated with very long lasting projects, such as Boeing's 777 airplane. Most virtual organizations will have shorter project lifetime expectancies.

FORMING A WEB

The group of organizations that contribute resources to the creation of virtual organizations are called a web. They have agreed to work together to form a virtual organization when an opportunity arises. The web is often thought of as an extension of the Japanese business cooperatives called Keiretsu. Typically, the companies have agreed to pool resources and are prepared to share people, information, and physical assets. They work out a way to enable the rapid and easy formation of an all-star virtual team because they have found that by combining resources, they can achieve more than any one of them could alone. A web of closely related companies organized to rapidly pool resources can create virtual organizations more powerful than a collection of arm's length companies with similar capabilities. Pooling does not mean co-bidding jobs, but actually integrating resource elements so that the whole is more than the sum of the parts, even though each company remains a separate legal entity.

> U.K. Fine Chemicals is a consortium consisting of eleven British fine chemicals manufacturers, most of which directly compete with one another. They formed the consortium as a means of creating a united marketing effort, primarily in the U.S. When they visit U.S. chemical manufacturers and association meetings and conventions, U.K. Fine Chemicals offers potential customers the combined competencies of eleven companies in any combination required to produce what a given customer wants. At the same time, the apparently greater scale of operations and scope of capabilities of the consortium vis-à-vis its individual member companies is also an advantage. The customer is buffered from the details of working out which of several capable companies will produce what.

The shared marketing relationships formed by U.K. Fine Chemicals brought them to the realization that they needed to form a virtual organization to meet the marketing opportunities they were uncovering. Similarly, the Agile Web of Pennsylvania, a group of small manufacturing companies, has found that it can collectively take responsibilities that exceed the capabilities of the combined companies. This capability comes from the ability to systematically integrate resources rather than use them as a collection of individual components. The Agile Web manages prototype development and new product launches as if it were the manufacturing arm of a design and marketing firm. As a collection of arm's length companies they could not take on that kind of responsibility.

Figure 16.2 is designed to illustrate how participants in a web come together to learn how to form a virtual organization. It indicates the close communication necessary to work with each other, and shows that more than one virtual organization can be fielded by a web at any one time. Each participant in the web does not have to take a role in every virtual organization. Resources, rather than whole companies, are combined through all-star teams set up to accomplish specific time-based opportunistic goals.

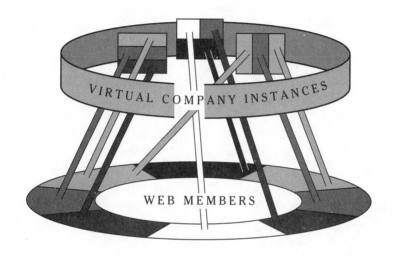

Figure 16.2 The Virtual Web Mechanism.

The idea of combining selected resources, as opposed to whole companies, reinforces the view of webs and virtual organizations as opportunistic with a strong need for mutual understanding and relationship building. Trust plays a major role in the ability of the web to succeed, and people are critical in facilitating this type of cooperation. The technological challenge, difficult as it is, is less of a hurdle for most webs than dealing with its people and culture.

THE WEB CHALLENGE

The first task in forming a web is to agree to the need for developing a shared vision, metrics, and goals. This most critical step is also the most overlooked. Once this is accomplished, the web faces the task of sharing ethics, values, and information, and developing communication using technology and associated methodologies.

The companies in the web should develop a trust and reward structure that they believe in, and that acknowledges the person-to-person nature of business relationships. Webs, like other businesses, must make a commitment to being world-class enablers of their customers, not arm's length suppliers. In an age of relationship-based competition, where price, quality, and delivery are no more than the entry tickets to competition, the web is a tool for building a virtual interprise with a significant depth of relationships and the ability to expand on them.

In the early days, when the benefits of cooperation were less understood, companies tended to join a web with a defensive attitude: they did not expect it to succeed, but wanted to be a part of it in case it did. This is a prescription for failure and wasting resources. Time has proved that a web can create all sorts of synergistic business opportunities if a few guidelines are followed. Select the best and brightest organizations to be your partners in the web. Play fair and seek win-win opportunities. Evaluate the situation in the context of the longer term value, not in the short term perspective of each fleeting opportunity. Motivate your people by sending the message that they should be both contributors and beneficiaries of the web and its activities, because the world is moving from arm's length relationships to interactive businesses. If you are not worth cooperating with, you will not be a desirable player in the evolving new business environment.

DYNAMIC, LINKED ORGANIZATIONS

The web affords the ability to dynamically create a virtual organization structure with attributes matched to the task or opportunity. If virtual organizations are to be time-based and opportunistic, they must also be dynamic and matched to the needs they are meeting. They can provide a virtual, vertical integration, using an integrated supplier network to utilize resources from around the world. The virtual organization concept was developed for its ability to pool resources to achieve world-class integrated engineering and design capabilities.

Networks now link various resources across business lines as they are needed to accomplish specific tasks. Interinterprise business and production systems with intensive, interactive, information sharing within and among companies are emerging to support internal teams and task forces, and to connect webs of suppliers and customers outside the organization—sometimes these will even include competitors. Virtual organizations use these capabilities to achieve concurrency and reduce design time as well as the overall concept-to-cash cycle time.

In many instances, the collection of organizations is first formed to offer the same one-stop shopping capability that its big competitors offer, as in the case of U.K. Fine Chemicals. The concept is constantly evolving, however, and large companies are now breaking themselves up into collections of smaller companies that act independently in well-defined niche areas of the original business. IBM, Alcoa, Asea Brown Boveri, and others have embraced the idea of having a collection of profit centers, each with a well-defined mission. They operate like a web of separate companies and pool resources to rapidly form virtual organizations to meet opportunities that cross divisional boundaries. In this way, the web has become a vehicle both for a collection of smaller companies who see a benefit in pooling resources, and for huge, multinational corporations to gain the corporate agility they need to compete in the new environment.

Early in its tenure in government, President Clinton's New Economic Council (NEC) staff sought advice on how they could reorganize the federal government to optimize their efforts. In discussions with us, they concluded that it did not matter what brilliant organizational structure they came up with, because no one structure would suit all purposes and be convenient for all tasks. Instead of trying to justify a major reorganization, they were advised to address each task or opportunity that crossed organizational lines by forming virtual teams. As a result, for several years the NEC has successfully used virtual organizations to meet its varied needs.

THE LEGAL FORM OF A VIRTUAL ORGANIZATION

When the idea of virtual organization became widely known, new legal forms were expected to follow. Research with precedent and expertise shows, however, that no new form of legal structure is necessary. Those forming webs and virtual organizations have used a broad variety of legal forms of organization. The range of legal solutions for virtual relationships is far greater than the well-used strategic alliances and partnering agreements commonly referred to. It is important to note that these agreements should not try to solve every imaginable dispute or misunderstanding, but instead should specify a method whereby unexpected problems can be dealt with. T*he Handbook of Virtual Organization* (Preiss, Leary, Jahn, eds. 1996), contains many worksheets and methods that can be used to deal with shared risk and reward, intellectual property, quality, liability, and the legal questions of virtual organization agreements. In addition there are examples of:

- how to negotiate agreements,
- clauses and texts for agreements,
- methods for demonstrating and managing quality and liability and warranty,
- dealing with intellectual property issues brought into the virtual relationship and created by it, and the policing of these against third party infringement,
- many methods of sharing risk and revenue equitably.

If a virtual relationship tries to rely on an agreement to survive, it will die. A viable virtual relationship starts with sound business sense. If there is a good business reason for the relationship, there will be an incentive to solve problems in a mutually suitable way. If there is no business reason in the background, no legal agreement can save the relationship. An open and honest sharing of views, expectations, ethics, and values is required, not to satisfy ideology, but to ensure bottom-line business success.

The idea of protecting and sharing intellectual property is intrinsic to the virtual relationship or web organization. With the inclusion of customers, suppliers, and even competitors in these sharing arrangements, a higher order of ethics and ethical behavior becomes important. There are no silver bullets or ironclad formulae for success, but honesty, ethics, and values are emerging as key enablers of virtual organization relationships.

As the competitive environment changes from product-centered, arm's length relationships to interactive and dynamic relationships, new modalities of cooperation are emerging. The legal, regulatory, and accounting solutions available may not yet provide special support or incentive for this, but they do not prevent this development. The precedents and methods described in *The Handbook of Virtual Organization* are not yet well-known, but they exist. As we have shown throughout this book, the move to the new cooperative, interactive world is an inexorable necessity, born of the constant search for efficiency in a competitive world. The enterprise has two choices: remain in the product-focused, arm's length, low profit environment, or move to the new competitive challenge of opportunistic relationships.

SUMMARY POINTS

⇨ A virtual organization is a new and evolving concept.

⇨ It is a natural extension of the enduring relationship-based competition.

⇨ It is time-based and opportunistic.

⇨ It is relationship-based and follows the Interprise Relationship Model.

⇨ The web is an organizational incubator for virtual organizations.

⇨ It is a way of learning how to build successful virtual organization relationships, using the Interprise Relationship Model.

⇨ The core competencies of a virtual organization are those of the web. Values and managerial ethics are critical issues to make explicit.

⇨ Webs and virtual organizations will succeed only when the people who create and work in them make them succeed.

THINGS TO THINK ABOUT

☐ Are the goals and objectives of the virtual relationship clear and action-oriented?

☐ Are the other participants committed to success?

☐ Are you comfortable with the ethics and culture that the web assumes and will deploy?

☐ Is an appropriate information infrastructure in place?

☐ Is the management attitude supportive?

☐ Do you have a culture of trust in place, where people can share goals, objectives, and constraints?

☐ Would others select you to be their partner?

CHAPTER 17

THE TRUST FACTOR

B ecause the interprise is based on multiple relationships outside the organization—whether with suppliers, partners, or
other stakeholders—mutual trust and shared values and
ethics are becoming a critical success factor in the new relationship-based business environment.

To be trustworthy is to behave in a predictable fashion, and to do
what you say you will do when you say you will do it.

Don Runkle, when he was vice president in charge of the
Saginaw steering division of General Motors, described
his organizational interprise relationships with the Japanese industrial company NSK like this:

1) NSK is our supplier; we buy supplies from NSK.
2) NSK is our customer; we sell supplies to NSK.
3) NSK is our partner; we have a strategic alliance or
   joint venture with NSK.
4) NSK is our competitor; we compete with NSK.

Runkle says that such a complicated set of relationships
would not have been possible in the past, but current
global business needs are driving both NSK and his Gen-

eral Motors division to find ways to achieve this multiple dimension interprise relationship. Success is clearly dependent on the ability of each company to trust the other.

In such a case, both parties to the interprise relationship must define in ways that are clear to both what is, and is not, acceptable and ethical behavior. Every one of the four relationships above carries its own set of ethical assumptions and behavior that should be made explicit if the relationship as a whole is to be successful. Each company must also address the problems that could arise when knowledge derived from one relationship is shared among the others. For instance, it might be unacceptable to use information learned from the partnering relationship in the competitive relationship. The interprises have to agree how information should be handled.

In their discussions, they will clarify the categories and classes of information and the levels of sharing and or protection they are comfortable with. This is not necessarily a call for lawyers to have a field day. Early experience suggests that a legal model that can be used on more than one occasion is better than a new set of agreements for each interprise relationship. This may not be necessary, however, if participants are willing to simply confer together, in order to discuss mutual expectations and create a set of ethical agreements between the interprise parties without involving lawyers. Simple language, spelling out in clear and uncertain terms what is expected of each party, is invaluable in either case. Most business leaders prefer clearly understood, unambiguous ethical statements to legal language they cannot decipher without legal experts.

Such agreements must cover not only acceptable behavior, and agreed methods of dealing with unpredictable situations, but also the sharing of intellectual property and proprietary information. For example, intellectual property brought to the table by party "A" may be restricted to use in that particular partnership effort, and otherwise remain strictly the property of party "A." Or the agreement might state that the intellectual property is to be licensed by party "A" for free use by party "B." The key to success is clearly to state the intent, and have all parties understand and agree to the processes for dealing with events as they occur. You do not have to anticipate or try to deal with every eventuality beforehand. This is usually an exercise in futility that wastes time and, worse, dampens enthusiasm and momentum for a project. Moreover, it is becoming clear that a reputation for keeping one's word is a major success factor for today's businesses and business leaders. In this environment, even the appearance of unethical behavior may be so costly in terms of lost interprise opportunities that iron-clad legal agreements will not be needed.

The successful interprise will not simply be a collection of organizations working together at arm's length. We see a growing trend toward resource pooling, and all-star interprise teams with specific, limited objectives. This will make it much more difficult to protect intellectual property by hiding it. Even with ethical agreements, the interprise will have to trust that participants from other organizations will respect their intellectual ownership. The alternative is not to become an interprise, thus forgoing all the business benefits that come with this new way of doing business.

INTERPRISE CORE COMPETENCIES

In an analysis of their core competencies, the Agile Web of Pennsylvania, a partnership of several manufacturing companies, discovered the importance of values and ethics in their business.

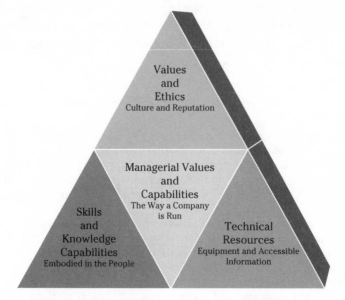

Figure 17.1 Analysis of Core Competencies.

In discussions with nineteen members of the organization, all agreed that in the pyramid shown above, the lower two triangles, representing skills and resources, were appropriate. Inclusion of the other two areas, "values and ethics" and "managerial values," surprised some of the members. While all agreed these were important, the idea that these were core competencies led to significant changes in the way the people at Agile Web viewed their organization. In the end, they decided that not only were the upper two areas core competencies, but they were to be the way the company distinguished itself from its competitors. They concluded that the managerial ethic of the Agile Web would be to provide more to its customers than they expected. Their new literature reflected this change of focus as you can see from the statement below.

In addition, they discovered that if any Web member had a skill or a resource, the whole interprise could claim to have that skill or resource. If you refer to the pyramid, you can see that this is a logical "or" connection relating the individual to the collective capabilities. In the central two triangles, however, the "and" relationship prevails. It is not enough that any one member of the interprise is honest. You cannot infer from that the entire interprise is honest. Thus, interprise members must not only police themselves, they must carefully screen new members, monitor themselves, and be watchful of the ethics and values of their partners. The interprise shares a common reputation, and one member can destroy what all must work to establish and maintain.

Agile Web Ethics Statement

The purpose of Agile Web, Inc. (the "Agile Web") is to bring several manufacturers together in such a way that, through collaboration and cooperation, we can bring higher value services to customers as well as obtain business that we could probably not get working as individual firms. This will increase our individual and collective competitiveness in the marketplace.

For the Agile Web to operate in an environment of trust and cooperation, and for all of its members to gain the benefits, economic and otherwise, of such collaboration, it is critical that all members subscribe to, and comply with, a common set of ethical standards. However, although an expression of values, this Ethics Statement is not legally binding and is not intended to create any legal or enforceable obligation of the signatory.

Thus, as CEO of (company name), a shareholder of Agile Web, Inc., my company subscribes to the following statement of ethics. We will:

...be trustworthy and honest in our dealings with our Agile Web partners. Our chief ethic is to be impeccably honest with other Agile Web members, our customers, our employees, and our suppliers. We recognize that our combined reputations are at stake with each and every Agile Web business opportunity. To insure our combined success, and to further develop trust within the group, we will never mislead any of our partners.

...keep our promises to our partners. We will treat our participation in Agile Web business with sufficient priority, giving it proper attention and balancing it with our regular business, to assure success. We will work with customers and partners to see that all are satisfied with the outcome. We expect to meet our commitments. We will honestly report to all involved, in an open and timely manner, any situation that arises which might impact the success of a project.

...commit to continuous improvement. Quality is defined by our customer, and is given in all that we do. Our organization is committed to a continuous improvement philosophy where we continually strive to reduce costs and improve response time, quality, productivity, and customer satisfaction.

...value our people. It is our people that provide the skills and knowledge required to serve our customers. We will strive to keep our employees informed, and allow them to grow and develop their skills so that they see themselves as part of the team.

...share information within the Agile Web that is necessary to get the best solution for our customers. We will encourage an open give and take of ideas in search of continuous improvement to our products and services. When necessary to develop the best solution for our customers, we will share information with other Web members, such as our costs (for a particular potential business opportunity only), current shop loading, changes in loading, our interest/need for the business, and so forth.

...hold confidential all information learned about our partners that is of a proprietary and sensitive nature. As we will learn information about our partners that they don't typically share, we will respect their confidence by treating such information as confidential, and by not disclosing or utilizing trade secrets or other sensitive information discovered through Agile Web activities.

...not compete with the Agile Web. We will not knowingly submit a proposal for business that is in competition with the Agile Web. Even if we leave and are no longer part of the Agile Web, we will not compete with the Agile Web on any business opportunities learned while we were part of the Agile Web.

...will respect and accept the decisions and consequences of the Agile Web President and/or the Agile Web Board. This is necessary for the Agile Web to respond in a quick and orderly way to our customers.

_____

(signed)

TRUST VALUES AND ETHICS INSIDE THE INTERPRISE

The most valuable resource in the interprise is its people. Their value derives from their skills, knowledge, and expertise, the information they can provide, and the profitable relationships they form. People are the most underutilized resource in today's businesses as is clearly evidenced by the "trust gap."

A trust gap is the difference between

1) The percentage of people in the organization who *can* be trusted to carry out goals and objectives given a set of constraints, and

2) The percentage of people in the organization who *are* trusted to carry out goals and objectives given a set of constraints.

In management surveys, most companies confess to a trust gap of between 20 to 60 percent with an average of about 40 percent. This suggests that most companies could significantly increase organizational effectiveness within the interprise with little or no investment in capital equipment. When asked why the trust gap is so high, company leaders blame inertia. In the pyramid structure of an old-fashioned organization, a few people at the top make decisions for everyone in the organization. As companies migrate to the concept of encouraging people to think and make decisions, there is a tendency to trust only as many people as you must to retain a sense of control. In addition, many seasoned employees are reluctant to be trusted because they have grown up with the old model, with managers who told them, Don't think, just do the job, we don't want your ideas. For them, doing what they are told is less risky than learning how to achieve a set of goals and objectives given a set of constraints. They are worried, Suppose I fail? Suppose the boss knows a better way? The drawing below dramatizes the positive impact of empowering employees.

Figure 17.2 The Power of Empowerment.

All management teams must seriously evaluate the trust gap in their organization, and, if a significant gap exists, take advantage of the opportunity to increase organizational effectiveness. Trust issues can occur between management and employees, among team members and whole teams. In most cases, a trust gap is an opportunity for empowerment. In some rare cases, there is a negative trust gap, where trust has been extended to those not yet mature enough to accept the responsibility of decision making. Usually, this means that further training and experienced-based learning are needed to develop the skills of the employees to be trusted.

To trust an employee is to confer responsibility on that person. If the consequences of this trust are business-as-usual or failure, the employee feels unfairly at risk. Trust and reward systems must be coupled to provide the right environment for the employees, and rewards should not only reflect success, but also prudent risk-taking. An organization that only values risk-takers who succeed will soon eliminate its risk-takers. Perhaps the greatest challenge is to find ways to define appropriate risk, and reward those who engage in it whether or not they succeed. Creating teams of risk-takers who are rewarded by the team's success is one way of accomplishing this. Leaders need to look for and reward behavior that reflects the value system they want to encourage; if they do this, they can trust that employees will soon learn and adapt to the new model.

In the old world of arm's length, stand-off, one-product-for-all relationships, emphasizing trust seemed quaint and even irrelevant.
In today's competitive arena, however, survival depends on becoming part of your customer's business or lifestyle processes. You cannot achieve this unless the customer can trust you. Similarly, as the
customer to your suppliers, you will not be able to gain the benefit
of their capabilities unless you are willing to trust them. When one
employee can harm the reputation of an entire organization, trust is
a people issue. It is not some touchy-feely discussion of old-fashioned values, but rather a substantive business reality.

SUMMARY POINTS

⇨ To be trustworthy is to behave predictably, and to keep your
   word.

⇨ Trust is a more significant competitive capability in the interacting interprise than it was in the product-focused enterprise.

⇨ Trust implies an accepted standard of ethics in and between interprises.

THINGS TO THINK ABOUT

☐ Is your business known to be trustworthy? Do customers trust
   you?

☐ What is the trust gap within your company?

☐ What is the trust gap with your suppliers?

☐ Do your people understand the full implications of not doing
   what they say will do on time?

Part 3

FROM ENTERPRISE
TO INTERPRISE

CHAPTER 18

TEAMS, NOT COMMITTEES

The interactivity that the interprise must necessarily develop with each customer individually is only possible when workers at all levels can interact freely. When interactions between businesses are conducted exclusively through executives, the businesses are not modern interprises. The paper *Industry Week* publishes an annual list of the ten best manufacturing companies in the U.S. Among the questions used to evaluate the companies, they ask what percent of the workforce can interact with customers. For the top ten companies, the answer is one hundred percent.

A team is relatively small—no more than a dozen people. It is a self-managed and empowered multifunctional group of people with a defined goal. Its members may be from different parts of one organization or from several different organizations. The team is able to deal with all aspects of a project. Teams are usually small to reduce communication problems and exclude members whose areas of responsibility are peripheral to the task. They are self-managing and empowered to act in such a way that delays caused by referring decisions back up the line are minimized.

The team is becoming the new work unit, just as the department was the work unit in the old system. A team is dynamic, flexible, and adaptive. It adds and spawns tasks, adding and subtracting people as needed. The department was an inflexible structure. Adding or subtracting people required a higher-level management decision which would often be preceded by a detailed study of the request.

> At the new rail car assembly plant of the PFA Planungs- und Produktionsgessellschaft Fur innovative Fahrzeu- gausstattung GmbH in Weiden, Germany, they abolished the specialization of work. All workers take an active part in the entire production process from the beginning of construction to completion of the InterRegio passenger cars of the German railways. The electrician sometimes hangs curtains, the mechanic lays cables. The workers organize the work and working time themselves and carry the main responsibility for the flow of material. One significant result is that the number of sick days lost has decreased dramatically.

Teams are not an end in themselves. They are a means by which the interprise interacts within itself, and with customers and suppliers. Teaming is justified only when it produces results. Putting teams in place without understanding why they are there and how they should be organized, can lead to disaster. The director of manufacturing at a large chemical enterprise described to us how he had been at a plant where a chemical reactor had blown up. No one was hurt. When he asked how the accident had happened, the plant manager told him that a self-empowered team had made the wrong decision. The director concluded that they should not have instituted teams. But the problem was not that teams had been put in place, it was that the team consensus had been allowed to override professional and safety considerations. A team facilitates interaction, but it cannot replace professional expertise. Alec Lengyel, who was for many years a senior manufacturing executive at General Electric, was twelve times a member of the company's trouble-shooting Tiger Teams, which were called in to fix problems caused by lack of professional expertise in previous teams.

Teaming is an important part of a restructured modern business, but because the management of teaming is often not understood, teaming sometimes leads to more problems than before. For instance, defining the new aim of the business, which is to become part of the customer's processes, is the responsibility of a leader. It cannot be left to a team.

A TEAM IS NOT A COMMITTEE

When a committee is established, efforts are made to ensure that the membership is balanced so that every constituency is represented, and no perspective neglected. Each committee member represents an interest, and the committee works together to try to find the best compromise between varied interests. In a true team, there are no sectarian interests, each person works to support the others in the pursuit of their mutual goals.

The fear motive in a company can be very powerful. Survival fears of an individual assigned to a team, especially in periods of change and downsizing, can be powerful enough to negate the effectiveness of teamwork. Employees work under the often severe pressure of personal economic survival that can get tied up with feelings of self-worth, affecting other relationships. This will turn the team into a committee of turf-protecting defensive participants. Successful companies promote positive teamwork, making a proactive effort to avoid defensive committee work. Given the right strategic focus and a reality of positive achievement, the team avoids becoming either a committee or a sterile debating society, but becomes a quick reacting, responsive key to company achievement.

Texas Instruments Defense and Electronic Systems Division, a $2 billion business that is becoming an interprise, understand this. As defense orders decreased, the approximately 10,000 people left in the division were rearranged into 1,900 self-empowered teams, each containing six to eight people, usually from different functions. The teams were given authority over their spending—as much as $10,000,000 for some teams, while many tens of thousands of dollars were spent by shop floor workers without obtaining managerial permission. Empowered teaming was a key factor in the successful reorganization of the division and contribution to the team was made an important factor in formal, periodic personal evaluations.

Teaming was an important factor in the continuing expansion of another company:

> In 1994, Solectron, the California-based, $1.5 billion manufacturer of electronic printed circuit boards and other subassemblies, and the winners of the 1991 Baldrige quality award, found that even though they were a proven high-quality organization, they needed to create empowered teams in order to continue to improve quality and delivery times even as customer requests became more volatile. ("It's Friday. The order for 1000 boards for next week I told you about. Change that to 200." "It's Wednesday. The order for 200 boards I told you about. It's now 2000, but you have more time. They are needed by Friday next week.") In their Milpitas plant, 88 percent of the Solectron workforce consists of minorities who speak fifteen languages, yet they succeeded in moving to empowered teams that enable them to continue to grow at a rate of 20 percent per year, in a competitive contract manufacture industry that operates at a small 3.5 percent margin.

COACHING AND EMPOWERING TEAMS

A team needs a coach. Steve Mills, human resource manager at Schrock Cabinet Company, says, "Coaches are the key to success in the team process. We missed that at the beginning." As a team deals with issues, the coach pays less attention to the issues and more to the group dynamic and method of personal interaction in the team. She or he takes note of the lack of teaming skills shown by any member, and sees that they get coaching to improve their team performance.

Companies often adopt teamwork in the wrong way and for the wrong reasons. Things aren't as good as they could be, so they try teams. A cross-departmental group is formed to deal with a problem. Because it is called a "team," the problem is expected to be solved. This does not work. The member of such a "team" pays allegiance to the person who affects his or her hiring, firing, and advancement, usually the department chief. The task on the team is then often understood as defensively protecting yourself and your department chief, not proactively advancing the company as a whole.

The interaction of the pay and reward system and teaming is a critical success factor that should be dealt with by experienced professionals. A balanced reward system motivates each person to improve the profit of the whole company.

> Toyota rewards managers not on the basis of the clever suggestions each manager has, but on the number of suggestions per year from the manager's subordinates, and on how quickly the manager responded to each subordinate's suggestion, thus promoting both leadership and teamwork.

> Mettler, the German producer of scales and weighing machines, found itself in the mid 1980s in a situation familiar to many companies. They were experiencing lower margins, and scrambling to deal with constantly changing customer requirements. The general manager correctly identified the cause when he said "the market refuses to adjust to our manufacturing system." As they moved to being a "make to order company," making only what had been ordered, then delivering it rapidly, Mettler organized the work by empowered cross functional teams. While other companies count and measure the number of suggestions per worker per year, and how these are dealt with, Mettler management does not know what suggestions are made. The teams deal with these improvements directly, and do not keep statistics on them. The only important matter is satisfying what each customer wants. Bonuses are based upon financial performance of the whole company.

To make decisions that advance company goals, teams need to be empowered. Empowerment, in turn, is enabled by information. People kept in the dark about company data will not be able to support company goals. An increasing number of companies are opening much information to workers, including periodic profit and loss statements.

> Springfield Remanufacturing, which rebuilds truck engines, was at the brink of failure in the 1980s. It then became an employee-owned company. Everybody in the plant has been trained to read the profit and loss statement that is posted regularly on the lunchroom bulletin board.

> NuCor Steel pays a weekly bonus based on the performance of the unit the person works with so that employees don't lose sight of the difference they make.

> Solectron taught all its people at the Milpitas plant to understand a profit and loss statement. Richard Seaman, vice-president says, "Managing a family is like managing a company. At home these people are presidents of their little companies. They can understand profit and loss." P & L statements are regularly posted in the lunch room.

TEAMING PEOPLE

Some personalities are more conducive to being a good team player, others less so. The Agility Forum in Bethlehem, Pennsylvania, has conducted extensive work on understanding the characteristics of people that support teamwork. Not everyone can have, or needs to have, all the skills needed to be part of a team, but taken together, the skill set of the people in the team should support positive teamwork. Personal skills divide conveniently into two categories:

1) independent skills, that do not require active interaction with other people
2) interdependent skills, needed to interact with other people

   1. Independent skills needed by a worker in an interprise include:

          mathematical reasoning; reading; writing; speaking; problem solving; time management; listening; presenting; synthesizing; analyzing; creativity; computer literacy; personal accountability; personal leadership; customer focus; risk taking; understanding of business metrics; self-direction; life-long learning; experimentation; reflective thinking; knowledge of systems; pattern detecting; innovation; cross-cultural competencies; quantifying measurement; awareness of workplace safety and health; understanding of workplace ethics

   2. Interdependent skills include:

          facilitation; communication; collaboration; persuasion; cross-functionality; negotiation; consensus building; understanding how others think and operate; trustworthiness; trust of others; teaming; role playing in a group dynamic; valuing diversity; leveraging complementary skills of others; conflict resolution; coaching; meeting management

A GUIDE TO SUCCESSFUL TEAMING

The basic rules for successful teamwork are no more than common sense applied to the workplace. It is almost trivial to write down the steps in organizing a team in a non-business setting, but these steps are seldom applied in business. For example, setting up a neighborhood sports team requires that you:

1) Decide which game to play—basketball, soccer, bowling, or whatever.
2) Choose a framework to train for—the little league, the annual city competition, or some other easily identifiable competitive sector.
3) Decide on the goal—winning the all-city competition, making it into the state finals, or another clear motivating aim.
4) Appoint a coach.
5) Only as the last step do you decide how to organize the team, and which players to recruit.

Many companies take the last step first, starting by appointing a team, and only then carrying out the preparatory steps. There are companies who decide to use teaming, get a facilitator to organize them, give the teams a pep talk, and expect magic to happen. There is no free lunch, and the unplanned approach to teamwork is unreliable.

The steps above can be systematically generalized into the following:

1) Decide on the strategic aim for the organization.
2) Work hard at projecting an unambiguous, clearly-understood, specific goal that generates enthusiasm.
3) Appoint a coach.
4) Align the method of evaluating each person so that the goal of the company becomes the goal of the individual. This not only promotes enthusiasm, but should assure the management that the individual will indeed align with the goal of the company.
5) Establish agreed-upon norms of behavior for the team.
6) Make sure success is well-defined, that there is real reward for success, and that winning or losing becomes each individual's concern.
7) Define the boundaries of authority, allow for negotiating for resources, and for changed boundaries.
8) Make sure that professional expertise is not overridden by the team consensus.

9) Reward entrepreneurship on the team, being careful not to favor the noisy extrovert over the quiet supporter.
10) Trust the team, letting them do things their way.

The key to all this lies in first defining the goal, then aligning the evaluation measures with that goal. Direction must come from the people who give the reward, which in most companies is management, since they deal with pay increases and firing. The goal should not be stated in vague general terms, such as "emphasize quality," or "reduce costs." It is important to carefully define the goal so that all the major issues are included, with room left for buy-in from the people in the team. Reward and performance is an issue to be approached with care. When an international technology company with high achieving workers introduced teaming, they arranged for bonuses to be linked to each team's achievement. What resulted was sabotage of each team by the others, in an attempt to increase the team's bonus by decreasing the other teams' shares.

TEAMS ARE HERE TO STAY

While there is a large and developing literature on *how* to team, it is important not to lose sight of *why* to team. Teaming is a means of interacting effectively with customers, suppliers, and within the company, enabling the interprise to adapt constantly with the aim of expanding the company for profit.

> Chief Executive Jan Carlsson defined SAS's goal as safety first, punctuality second, other customer service aspects third. Anyone in the organization who knows this definition has a framework for their decisions. A manager knows that safety comes before a punctual take-off, but a flight is not held up while waiting for extra blankets.

> Empowerment of individuals in teams has led to rapid improvements at Volkswagen. Otto Timme, director of the car component manufacture department of the plant in Braunschweig told his local paper, "Things are progressing at enormous speed. In the past, there used to be some

kind of weariness. To have a single machine retooled, it
was necessary to write lengthy letters to the board of directors. Empowered teams analyzed the production
process within five days, never forgetting the goal, and
quickly put their ideas into practice. In the piston pre-
assembly, the number of rod batches produced per
worker is now 222, compared to 133 in the past; the increase in productivity is 67 percent. The reject rate used
to be 305 per month; now it has gone down to 195. Today,
700 engines are produced per shift, instead of 500 as in
the past, although working time has decreased from 416
to 374 minutes." (Harold Schultz in the daily newspaper
Hannoversche Allgemeine Zeitung, 7/7/1993.)

Empowered teams enabled the BMW plant in Regensberg
to reduce tangible assets by 33 percent, and increase productivity 50 percent while keeping capital resources constant. Operating time of machinery increased to 99 hours
per week, saving between 300 and 400 million DM.

Teams clearly bring much benefit to companies, but successful
teaming applied to the wrong strategic goal, will not save a company. The business can be wonderfully efficient, but it can be doing
the wrong things. It is fascinating how difficult it can be for a team
to agree on a goal. Left to itself, even a positive team cannot be expected to arrive at the correct strategic aim. The context for the
team work is the responsibility of management, not the team. The
goal for the interprise is become part of the customer's processes.

The goal of becoming part of a customer's process requires that
the internal structure of an interprise become adaptive, populated
by motivated and knowledgeable entrepreneurs. This requires
teams. Teams are not a passing fad, but a central part of an interprise, replacing the old rigid hierarchy. They are here to stay, a permanent part of the new business scene.

SUMMARY POINTS

⇨ The interprise requires a rapid adaptability that is best served by teams.

⇨ Teams are not an end, they are a means to achieve profitable interactivity with customers.

⇨ Teaming within the company extends to customers, suppliers, and partners.

⇨ A team needs to be coached and empowered.

⇨ Teams members require different skill sets than they did in the old organization.

THINGS TO THINK ABOUT

☐ Does the method of pay and bonuses in your company support teamwork by the individual?

☐ Does the method of rewarding teams encourage teams to support the companies overall success, or to sabotage each other's efforts?

☐ Do employees of the company understand the goal of the company well enough for them to take initiative? If not, what can be done about it?

☐ Are employees in the company scared to make mistakes? If so, is this good or bad?

☐ What percent of people's time in the company is spent thinking, compared to doing what they were told? If this number is as little as 5 percent, as in many companies, what can be done to harness the wasted thought-power?

CHAPTER 19

FOLLOW
THE MONEY

Don Runkle, vice president of General Motors, has pondered long and hard over the reasons why the Japanese spent decades developing the lean production system, while the U.S. persisted with less efficient management methods. After all, the U.S. is the world's leader in technical research and development. He concluded (correctly) that the major factor which caused the U.S. to lag behind was its accepted method of cost accounting. Cost accounting is much more than a management tool; it is an empowerment tool. Before an entrepreneurial-minded person brings a suggestion to the team or the management, he or she first checks to see if the idea will bring a bottom-line financial improvement. If not, the idea never sees the light of day. Cost accounting has killed many good ideas, because it is fatally flawed.

Although many companies continue to use the erroneous cost accounting method for management decisions, an increasing number of businesses are introducing new methods. This chapter discusses the most important of these new methods: Activity-based Costing, Target Costing, and Time-based Costing. Of these, Time-based Costing has the most potential to provide suitable criteria for managing an interprise.

"192 COOPERATE TO COMPETE"

Generally Accepted Accounting Principles ("Peanut Butter" Method)

Activity-based Costing

Target Costing

Money Flow ("Throughput") Costing

Figure 19.1 Four Cost Accounting Methods.

In 1995, a group of small manufacturers located in eastern Pennsylvania, decided to work together as a virtual organization to bid for jobs which none could get on their own. The first few bids they submitted jointly were all rejected as too high. After some investigation, they found that the costing method they had employed was inappropriate. Each company used the commonly accepted costing method, but with different overhead formulae. This led to implicit cross-subsidization between companies that was not at all apparent initially. Not only had they been using the accepted (and erroneous) costing method, but they had compounded their problem by using it differently in each company. This produced estimated costs that appeared higher than they actually would be.

WHY THE TRADITIONAL COSTING METHOD IS WRONG

Most businesses use the accepted "peanut butter" costing method, in which indirect or overhead costs are spread across the direct cost of product or service. This accounting often leads companies to abandon profitable products or to pursue unprofitable ones, because the method of allocation suffers from a fundamental theoretical error. It is wrong. In their well-known book *Relevance Lost*, Kaplan and Johnson, two leading modern authorities, say the following about current accounting procedures:

Today's management accounting information, driven by the procedures and cycle of the organization's financial reporting system, is too late, too aggregated, and too distorted to be relevant for managers' planning and control decisions. [This system can lead to] misguided decisions on product pricing, product sourcing, product mix, and responses to rival products.

This is a serious indictment. How could sensible people let such a situation develop? It started in the late nineteenth century, when managers of many metal-working companies developed formulae to accumulate and analyze the costs and efficiencies for converting raw material into product. They developed standard measures and standard rates in order to plan the most efficient work flow for a factory. Costs were divided into direct labor costs and indirect costs, with the indirect costs—today called overhead—allocated to each product in proportion to the amount of direct labor invested in it. This method was applied in factories which produced few products, and where the overhead resources were used by each product at about the same rate.

When the accountants adopted the formula for general use in organizations as diverse as manufacturing and retail distribution, they neglected the warnings of people such as Church, who published the following around a hundred years ago:

"... it is very usual practice to average this large class of [overhead] expense, and to express its incidence by a simple percentage either upon wages or upon time ... As a guide to actual profitableness of particular classes of work it is valueless and even dangerous ... in the case of a machine shop with machines all of a size and kind, performing practically identical operations by means of a fairly average wage rate, it is not alarmingly incorrect. If however, we apply this method to a shop in which large and small machines, highly paid and cheap labor, heavy castings and small parts, are all in operation together, then the result, unless measures are taken to supplement it, is no longer trustworthy."

Holden Evans, a naval contractor around the turn of the century, described how even then cost accounting practices were driving businesses to difficulties:

"In some of the large establishments with numerous shops the expense burden is averaged and applied on the basis of direct labor—notwithstanding the fact that in one shop the shop expense percentage is nearly a hundred while in another it is less than twenty-five. Frequently such establishments are called on to bid for work which is almost exclusively confined to the shops where the expense is low, and by using the higher average rate the bids are high and the work goes to other establishments where costs are more accurately determined. Thus profitable work is often lost."

The identical source of error persists today. Many businesses, large and small, are equally misled every day by traditional costing methods. They give up profitable products or emphasize losing ones, because of the cost formula, not because of the realities of the costs. The inventors of cost accounting methods knew that they could be applied only to similar operations in single-activity companies, yet we continue to apply them today without realizing that they are so flawed as to be dangerous.

HOW BUSINESSES TODAY USE TRADITIONAL COST ACCOUNTING

Nick Franklin, a senior engineering manager at General Motors, is not confused by the costing method. He once installed a simple electric drill press which the accounting system combined with an expensive automated machine. The hourly rate on the drill press was very high since the drill had to carry the allocated cost of the automated machine. Franklin's solution was simple: he had a yellow line painted on the floor around the drill press, dividing it from the expensive machine, then separated it in the accounting system by adding it instead to the office equipment column. The overhead rate went down remarkably, pulling down the hourly cost of the drill press to a reasonable amount! He has had to do this many times, and calls it "the paint method of cost accounting." Standard costing methods force sensible people to resort to management contortions to get things done. Many, maybe most, managers work, not to make the most money for their companies, but to satisfy accounting formulae.

Companies certainly know how much total revenue and expense they have, but they do not know how much the cost of a particular product or project is. The confidence of an executive who says, "The cost of this product is too high. Reduce it!" is misplaced. The statement would be more appropriately phrased, "According to our cost formulae that are known to be inaccurate, the cost of this product is too high." There are many tragic cases of plants which have been closed, people dismissed, and profits abandoned, all due to the blind application of cost formulae.

Despite the many people who have seen the defects in cost allocation as it evolved over the years, and despite the many different cost accounting methods that have been suggested to deal with it, it still remains the method used with the "Generally Accepted Accounting Principles" (GAAP). All indirect costs are allocated proportionally to direct costs without analysis of the real impact of these numbers.

The example below is adapted from a real case, and shows how this method of allocation can predict a low margin on a profitable product. In this case, the European holding company of an Ohio manufacturer had specified that each project overcome a 50 percent gross margin hurdle. If the U.S. company had not stuck to its guns, they would have passed up a profitable opportunity, and suffered the consequences of reduced annual profit.

A product was proposed at a manufacturing company with a list price of $250. The net sales price to the manufacturer is $150, and the annual sales would be 140,000 units.

Unit costs are predicted to be:
    Material: $90 per unit
    Direct labor: $4 per unit

The standard cost structure for the company is:

Labor	6%
Manufacturing overhead	24%
Material and supplies	70%

Every dollar spent on direct labor is thought to incur four dollars of manufacturing overhead (24% / 6% = 4).

For the proposed project, projected expenses are:

One time:
    Design and development $500,000
    Product setup and preparation $2,500,000

Annual:
    Marketing and sales $3,000,000
    Manufacturing administration
      and planning $1,000,000
    Indirect freight and other
      costs of goods sold $2,000,000

On the basis of the above data, the cost per unit is assumed to be:

Material	$ 90
Direct labor, which is not hired and fired specially for this project	$ 4
Manufacturing overhead = 4 x 4	$ 16
TOTAL	$110 per unit

Gross margin = (sales price - manufacturing costs)/sales price

= (150 - 110)/150
= 40 / 150
= 27%

Based on the European company's 50 percent gross margin hurdle rate, the project should have been rejected.

However, this project shows real bottom line profit, as follows:

Predicted sales income = 140,000 x 150 = $21,000,000

Variable cost   = 140,000 x 90 = $12,600,000

Real income per year = 21,000,000 - 12,600,000 = $8,400,000

Additional annual expense =   3,000,000
                                           1,000,000
                                           2,000,000
                                          $6,000,000

Annual profit   =   8,400,000

                                 =   - 6,000,000

                                      $2,400,000

One time cost = $3,000,000

Payback time   =   3,000,000/2,400,000 = 1.25 years

Sensitivity analysis shows that:

- If the one-time costs increase by 50 percent, the payback time increases only to 1.875 years.
- If the project went so slowly that the annual sales volume and the accompanying sales, administration, and manufacturing effort turned out to be half of that predicted, the payback time would be 2.5 years.

This is a profitable project, which would have been rejected by GAAP. In fact, the Ohio company had a very difficult time because it correctly insisted on accepting the project, which was in fact very profitable, while its European owner looked only at gross margin.

Modern managers run their companies "by the numbers." But the numbers are fatally flawed! They lead managers to reject profitable projects, and to emphasize unprofitable ones. The adverse effects of the arithmetic error in conventional cost accounting are serious, and companies big and small are increasingly turning to other methods.

ACTIVITY-BASED COSTING

Activity-based Costing (ABC) developed from the understanding that overhead expenses should not be arbitrarily spread over projects, but should be specifically associated with the projects incurring those expenses. Expenses for activities are incurred when a series of processes is performed. Costs are incurred by using material and by executing activities. The cost of the activities is not always obvious when using the regular GAAP costing method.

In the early 1990s, a North American company engaged in manufacturing hospital products ventured into a new product line which required a unique package design and a unique combination of components for each lot. According to traditional accounting methods, this was a very profitable line, as compared to the marginal profits of the standard product, although they noted that no competitor had gone in that direction. Sales for the specialty product grew rapidly, but overall profits continued to decrease. Before implementing a decision to abandon the standard product, and concentrate even more heavily on the specialty product, they made an ABC analysis and discovered that the cost of the standard product line had been overestimated and that of the special line underestimated. The distortions of the regular costing method were hiding the fact that the standard product line was subsidizing the specialty line.

In ABC, costs are allocated to both material and to activities. Activities are performed by resources, and the use of those resources costs money. The stages of ABC are as follows:

1) Identify the major process activity categories, often called process mapping or transaction analysis. Examples of activities may be machining, inspection, or locating and delivering filed documents in an archive.
2) Reorganize the general ledger so that the line items are grouped according to the activity categories.
3) Allocate the total costs to the activity categories by interviews and time measurement.

4) Identify the cost drivers of each activity category. Drivers include square feet for office space, head-count for salaries, and number of accesses for archive retrieval.

5) For each activity category, compute the driver rate. Examples are dollars per square foot, dollars per person, or dollars per archive access.

6) To each project or product, assign the proportion of the activity cost used by it. The equation for this is the number of activities used multiplied by the driver rate. If there are twenty uses of the "locate and send file" driver in a project, and the driver rate is $100 per use, the project would use 20 x $100 = $2000 worth of that driver.

7) Add the costs of activities and the costs of material for a project or product to yield the total cost.

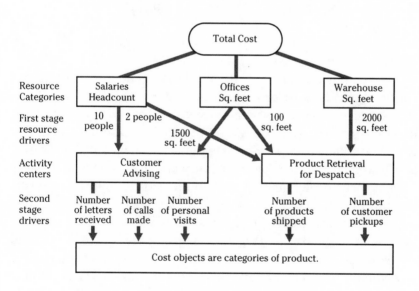

Figure 19.2 Activity-based Costing.

Some companies have used ABC with great success, but it has not been widely adopted. One reason for this is that analysis of all the activities in a company is an expensive, time-consuming, and possibly disruptive exercise, which can involve the need for a large, expensive computer system. Another reason is that there is an element of arbitrariness in the allocation method that may lead to excessive inaccuracy. Furthermore, if the structure of the company changes over time, the selected categories need to be reviewed.

There are many different approaches to implementation of ABC, depending on the structure of the company. The 80/20 rule of thumb applies here too: approximately 80 percent of the expense is generated by 20 percent of the activities. It is therefore usually worthwhile to pay attention only to the approximately 20 percent of activities in a business unit which generate the most expense.

TARGET COSTING

If the heart of interprise strategy is becoming part of the customer's work processes or lifestyle, the target costing method sets that aim as its goal and uses it as a guide to plan work that ends up as product and service for the customer.

The customer will pay for the total system and its features, not caring about the cost or performance of a component or subsystem. Target costing starts with a very careful evaluation of what features customers would be willing to pay for, and how much they would be willing to pay for each feature. Then, working backwards, an allowed cost is calculated for every action or item which makes up that cost—dealer margins, distribution costs, importation costs, assembly costs, and so on, down to the allowed cost for each component. The target cost for the system remains fixed. If a component cost exceeds its target, it is returned to research and development for further work. If it turns out that the cost target for any component could not even be met with more research and development work, the cost of some other component will need to go down. When a component does match its allowed value, further cost decreases are sought from it so as to allow margins of increase elsewhere.

When the Olympus Optical Company adopted this approach in 1987, only 20 percent of proposed new camera models cleared the cost hurdle in the first pass. The other 80 percent were not abandoned. The product development team continued to investigate whether the features offered could justify a higher price. In evaluating costs, the whole life-cycle of a product or service was considered, so that the cost of a development which could serve several product families did not need to be totally repaid from the first model sold.

Target costing forces close coordination with

- customers, to find out in great detail what price can be asked for which functions,
- departments and individuals in a company, because the trade-offs require constant analysis and interaction, and
- suppliers, since their input contributes up to 80 percent of the total cost.

This method also provides a strong guide for design and planning efforts. For instance, at Olympus, the pressure of target cost forced the number of parts in the shutter unit of one class of compact cameras to fall from 105 to 56, decreasing the cost of this subsystem to 42 percent of its initial estimate.

The target costing method is applicable to any business, from product assembly manufacturing such as cars, cameras, bulldozers, and electronic equipment, to processing, where the variables may be temperature, pressure, time, and purity of raw material, and to services where the items considered may be time for an activity, information processing, data, or mailing and distribution efforts. As enterprises migrate to becoming interprises, the move from cost-plus-margin costing to target costing naturally follows.

THE MONEY FLOW METHOD

The money flow method for computing costs is based on rigorous mathematics, and also matches common-sense understanding. It is

being successfully applied by a growing number of diverse companies. This method does not allocate overhead costs to projects and
products. Rather, the rationale is that money is invested in a company in order to provide a good return over time:

For a product

$$\frac{\text{Income}}{\text{month}} = \frac{\text{Income}}{\text{unit}} \quad \text{x} \quad \frac{\text{Units sold}}{\text{month}}$$

By focusing on margin, which is income/unit, we implicitly as-
sume that the velocity of product through the system—units sold
per month—is constant and does not affect the income per month.
That used to be true in the old world of slow and isolated arm's
length processes, but is not true in the dynamic, interactive world
of today's interprise.

This is illustrated by a simple example. Given the data below,
which is the more profitable product?

	Product A	Product B
Cost	$50	$50
Sales Price	$100	$80
Contribution Margin	$50	$30
Contribution Margin %	50%	37.5%

The contribution margin of A is 50 percent compared
with 37.5 percent for B, and most people would conclude,
based on the wrong criterion of contribution margin, that
A is the best product. The correct answer is that there is
not enough data to make any such conclusion.

The times required to produce the products are:

	Product A	Product B
Time in hours	12	3
Contribution margin $/hour	50/12 = 4.17	30/3 = 10

At this point it becomes clear that B makes more than twice as much money per hour as A. The simple calculation above helped a particular director of manufacturing understand why the competitor was making much more money even though his margins were higher than those of his competitor. The competitor was looking at how much money was made per hour, while he was looking at the margin per unit of product.

Bearing in mind that work is achieved by processes, the money flow method starts by charting the flow of work through processes. The following terms are commonly used in this method:

- The *workpiece* is a unit of product, a chunk of information, a paper file, a letter or form being worked on, or any other item which is processed.
- A *resource* is a machine, a person, a department, or any other element which acts upon a workpiece.

Figure 19.3 shows the flow of workpieces through a company, where the resources are shown as blocks.

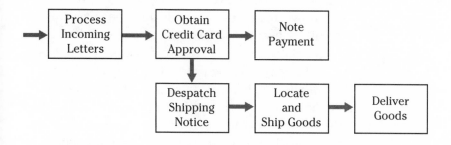

Figure 19.3 A Simple Process Map.

By following the flow of workpieces, you can follow the flow of money, and by a methodology which is surprisingly simple, get an accurate picture of the cost of a product or project. Rather than thinking of money as tied to a product or project in a company, money values are thought of as a flow—dollars per hour, day, week, or month. In this explanation we use the month as the unit of time, but any unit of time can be used.

Running a company requires the measurement of two kinds of money flow:

1) *Variable* money flow is the variable monthly rate, the actual dollars per month flowing out, and actual dollars per month flowing in from sales, for each product or project undertaken. These sums include no allocations or loadings, but are the real flows of money actually incurred or coming in. The variable flow is computed for each product or project.

2) *Constant* money flow is the cost of all the other activities which are not part of the variable flows, but keep the company going. The constant flow is computed for the business unit as a whole. This flow is said to be constant because it does not vary with the daily sales of products. As companies downsize or expand, the change in the flow of money which is not tied to any specific product or project depends on overall management decisions, not on the specific details of each month's sales income and variable costs. In the nineteenth century, when the old costing method was invented, it was common for workers to be paid for each item they made, and to bring their own tools to work. In those days, overall expenses were indeed closely tied to the output of goods, but this is not the case today when people are paid by day, week, or month, and the employer provides the tools and machines.

The practical way to determine the variable cost for a project or product is as follows: If you are making a product or managing a project, imagine stopping the production or project, and write down all the costs which would be saved. That decrease is the variable cost. It includes purchased material only if the order can be stopped, the people only if they would be sent away, and other costs only if they indeed will be saved. If you have yet to start the project or make the product, variable cost is the increase in dollars per month which would be incurred when running it. This leads us to a few simple definitions:

- The *financial throughput (T)* of a company is the rate of dollar income per month less the rate of variable cost per month:

    Financial throughput per month (T) = Sales income per month - variable cost per month

- The *operating expense* is the constant dollar flow rate per month needed to keep the company going. It includes every expense which is not directly part of a variable cost anywhere in the company.

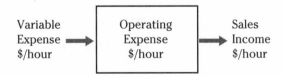

Figure 19.4 The Money Flow.

The money flow is related to the sales income and variable cost for each unit of product, by the following simple relationship. For each unit of product or service,

Contribution Margin = Sales income - variable cost

The financial throughput per month for the business unit T, is the sum of the contributions of each product or project to the throughput, i.e.

Financial throughput per month (T) =

(Contribution Margin/Unit) * (Units/month) for project 1

+ (Contribution Margin /Unit) * (Units/month) for project 2,

+ (Contribution Margin /Unit) * (Units/month) for project 3, etc.

When using the contribution margin method as a criterion for choosing a product or project, it is assumed that the flow rate of units through the system has no effect on the financial return achieved. This approximation was reasonable in the early days of mass production, but it is erroneous today. The money flow method uses the financial throughput rate (T) as the decision criterion. To do this, we need to investigate whether or not there are limits to that flow rate.

When following the flow of money through a company, we should be aware of the importance of bottlenecks or constraints on that flow rate. This is demonstrated by the following example.

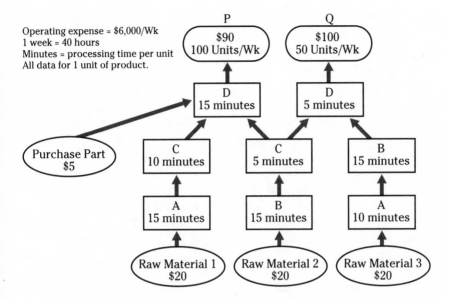

Source the Haystack Syndrome, page 67

Figure 19.5 Money Flowing through a Company.

From the information in the figure, the contributions margins are:

- for P, $ 90 - $45 = $45 per unit
- for Q, $100 - $40 = $60 per unit

If we produce the full market demand for P and Q, the income per week will be:

- for P, 100 x $45 = $4500
- for Q, 50 x $60 = $3000
  Total $7500

The profit for the week then appears to be $7500 - $6000 = $1500

This is the result that would be given by any system which neglects finite capacities of resources or the flow rates of money. In the following chart, note the times needed from each resource shown in Figure 19.5.

RESOURCE	A	B	C	D
minutes/unit of P	15	15	15	15
minutes/unit of Q	10	30	5	5
minutes/100 units of P	1500	1500	1500	1500
minutes/50 units of Q	500	1500	250	250
Total minutes required	2000	3000	1750	1750

One 40-hour week has 2400 minutes, but we need 3000 minutes of resource B, so the market requirement cannot be met. If we prefer product Q because it has a better margin, we will first make all 50 units of Q in 1500 minutes; then, with the remaining 900 minutes of resource B left over, we will produce 900/15 = 60 units of P.

The weekly financial throughput is then:

Q   50 x 60 = $3000

P   60 x 45 = $2700

$5700

and the weekly profit is $5700 - $6000 = ($300), a loss of $300.

Relying on the wrong margin criterion, the business appears to be losing money. The criterion should select the product which produces the most income per unit of time, not per unit of product.

To calculate the money flow criterion for each product, we need to know the time taken to produce a unit of product. The net income divided by the time would give the money flow. If the system had no bottleneck we could use the total clock time needed to produce the product. However, if the system does have a bottleneck, we should choose the time taken at the bottleneck, to make the correct choice. This is a well-known result from the mathematics of flow networks, explained and popularized by Eli

Goldratt in his books *The Goal* and *The Haystack Syndrome*. In this case, there is a bottleneck, resource B, so we need to choose the product which makes most money per unit of time at the bottleneck. The flow rates of money through the bottleneck are:

- for a unit of P, $45/15 = $3 per minute
- for a unit of Q, $60/30 = $2 per minute

Product P produces more money per unit time, and should be preferred. If we produce all the market demand of 100 units of P, this will require 1,500 minutes, leaving 900 minutes to produce what we can of Q. This will be 900/30 = 30 units of Q. The weekly throughput will then be:

- for P, 100 x $45 = $4500
- for Q,   30 x $60 = <u>$1800</u>
                        $6300

This will yield a profit of $300 above the weekly operating expense of $6000. The product with the lower margin gave the higher profit!

The lesson to be learned from this is to choose projects and products based on the dollars per day they produce, not the dollars per unit of product or service. This is a profound result that points up the error in today's common business practice of choosing projects or products based on the contribution margin for a unit of product. The criterion of money per unit time was always the correct one. As accounting methods developed, they approximated the criterion of money per unit of time by the factor of money per unit of product. In the old days of arm's length slow enterprises, which specialized in few products or projects, you would double the income per unit of time if you doubled the margin per unit of product. In today's interprises, which deal with very many rapid and specialized demands from specific customers, this approach leads you astray.

TIME-BASED COSTING

Using the money flows for income and expense in a company is the correct approach, both mathematically and instinctively. Ask someone, "Is a million dollars a large sum of money to earn?" A teenager inevitably says "yes," but the answer depends on the time period involved. A million dollars a year is a large salary, but a million dollars over the forty years of one's working life, at $25,000 per year, is relatively low. Income and expenses should be thought of as flows, not as dollar sums. Operational decisions as to which products to choose and which investments to prefer should be based on allocating money to units of time, not to units of product.

Costing methods are used to prioritize management decisions. Time-based costing uses the money generated per unit time as the prioritizing parameter. This method is discussed in more detail in our previous book, *Agile Competitors and Virtual Organizations: Strategies for Enriching the Customers*, and the steps in implementing it are given below. When using this method, companies often find that the data available includes many errors. The critical data must be checked for accuracy; non-critical data, which is usually most of it, will not affect operational decisions.

The time-based costing method works as follows:

1) Decide which functions are included in the business unit to be analyzed and which are excluded.
2) Draw a diagram of the flow of workpieces through the resources in the business unit (similar to the preceding PQ diagram).
3) For each product or service:
   3.1) Find the sales price and the truly variable expense.
   3.2) Calculate and find whether there is a bottleneck internal to the process, when dealing with the total project and product mix the organization deals with. This is done by calculating the time needed at each resource as in the table above, to see if any resource would be required for more time than is available.

3.2.1) If yes, compute the financial throughput dollars per time of constraining resource, for each product, and prioritize them based on this.

3.2.2) If no, note the money flow which is the financial throughput of dollars per total time required to get a unit through the system. Since there is no internal bottleneck, the internal system can take in more work; the bottleneck is in marketing. Prioritization should aim at increasing marketing efforts.

4) To get the total income, add up the results of all the products or services in 3) above.

Figure 19.6 is an analysis of products made by a plant of the Rhône-Poulenc corporation. It shows the contribution margin for each product and the net flow of dollars per hour contributed. It is obvious from looking at the chart that by using the contribution margin criterion, the sales force was pushing less worthwhile products, and some worthwhile products were being downplayed by the sales force.

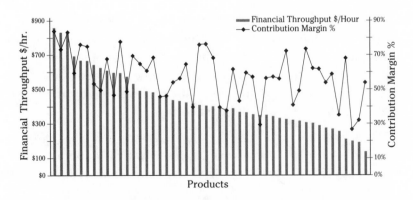

Figure 19.6 Financial Throughput Gives a Different Prioritization of Products than Does Contribution Margin.

The dynamic, time-dependent nature of the environment in which the interprise finds itself was described in chapter 3. In this environment, time rates are important, and the financial benefit of a product or project must be evaluated by looking at the money made per unit time, not per unit of product. The errors due to using contribution margin as a prioritization measure are compounded by the fact that production runs are becoming shorter as competition forces companies toward individualized products and services and fragmenting markets. The use of the money flow criterion is a profoundly different approach than that adopted by most businesses, but experience in those companies which do use it, finds that it brings them real bottom-line success. This method can coexist with the financial reporting methods for the Securities Exchange Commission and other external agencies and interested parties. For the interprise, it will probably become the preferred method of management cost accounting. In the meantime, the key is to use time rate measures rather than absolute values.

SUMMARY POINTS

⇨ The Generally Accepted Accounting Principles (GAAP) are satisfactory for periodic external reports, but the use of contribution or gross margin is the wrong criterion for the interprise to use to prioritize and cost projects.

⇨ Activity-based Costing is used to allocate overhead to the projects or products using it, but it involves arbitrary decisions. Introducing Activity-based Costing can be an expensive process, and you need to remember that 20 percent of activities often produce 80 percent of the cost. This enables one to carry out a limited, but useful, activity cost survey.

⇨ Target costing starts with the customer and allocates allowed cost to each stage of the work. A project which cannot reach its cost targets is either reorganized completely until it does, or rejected.

⇨ Time-based costing uses the more realistic criterion of net revenue per unit time from a product or project. In applying it, one must be careful to evaluate the influence of possible constraints on the flow of workpieces through resources.

THINGS TO THINK ABOUT

☐ Are you selling many products which have good margins, but overall annual profitability appears inexplicably low?

☐ Are you selling customized services or products at prices much lower than the competition? If so, why aren't they matching your prices?

☐ Do you compare cost margins only with your competitors, or do you compare the flow rate of money produced?

CHAPTER 20

THE PEOPLE
WEAPON

Some people feel that there is an implicit contradiction between the importance of teaming, and the value of individuals and individual contributions. They feel that a supportive team member must necessarily inhibit personal ambition. This is the result of an incomplete view of teaming, and of the natural tendency of individuals to compete. As teams are used more and more, and as leadership style changes from dictating to coaching, the individual becomes even more critical to the success of the organization. Paradoxically, just as the fiercely competitive business climate is pushing companies to value employees for their motivated entrepreneurship rather than their obedient muscle power, the same climate leads them to shed employees in brutal and often misguided downsizing efforts.

Asea Brown Boveri has over 1400 teams deployed throughout the world. CNN also uses global self-motivated teams. More and more organizations are realizing that the solution to the morale problem is to empower their employees in this way. As workers, they expected the company to take care of them. When the company could not promise them the security they were accustomed to, they lost their motivation. Organizations can alleviate this distress by putting employees in charge of their own destiny. Working with people to

enhance their competence and skills makes them the best employees they can be. If companies help their workers increase their
worth, they will increase the company's worth even more. And, if
outside forces mean they have to leave, they will be trained and prepared to move on.

PEOPLE ARE ASSETS

The thesis of this book, derived from much observation of modern
companies and many interactions with company leaders, is that it is
possible for a company to expand while the corporate landscape
undergoes its current epochal change. Successful companies are
coordinating changes in the way they work with customers and suppliers, as well as in their internal restructuring. Unfortunately, many
other companies have not yet realized the interaction and synergy
between these three areas. They deal only with internal efficiencies,
assuming that the products they have always made and the relationships they have had with customers and suppliers will remain
constant. So they cut people, gaining short-term benefit according
to financial formulae, but failing to cash in on long-term improvement. By losing the knowledge and experience built up by people
over time, companies eliminate the seed corn for future growth;
they do not recognize that in the new environment, people are a
critical asset for any organization.

In the old days, a successful company wanted to go it alone, not
share its success. It was vertically integrated and needed no outside assistance to excel in its field. This is no longer possible for
most organizations. Companies, of course, need to add and sub-
tract people from time to time, but the way it is done now, with blan-
ket decisions to eliminate a given number of people without
detailed analysis of customer relationships and core competencies
which need to be nurtured, is an unfortunate consequence of per-
ceiving machines and buildings as assets and people as expenses.
Professor Dan Roos of MIT echoed the thoughts of many when he
said in a public speech in 1994 that in a modern industry, machines
should be treated as expenses and people as assets.

In the old mass production system which existed when current
accounting practices were first set up, machines were expensive

and usually custom built, designed to be used by uneducated people who were hired and fired at short notice. Today's circumstances are quite different. Capability is governed by the complexity of the value-adding system as a whole, not the difficulty of making a particular component. Most machines and machine capacity are widely available for hire from suppliers, but it is the intellectual capacity of the people who operate the entire competitive system—from interacting with customers, through planning and designing, to implementation and delivery and support of the customer—that drives the competitive engine.

In chapter 4, and in many examples throughout the book, we have emphasized the need for speed in today's economy. Speed alone, however, without the ability to accelerate and decelerate, is useless. The usual method of accelerating and decelerating, derived from the traditions of the vertically integrated, capital intensive industry was to hire and fire. In today's competitive environment, this method is clumsy, and flies in the face of the need to nurture the motivation and knowledge of people, so that the total value-adding system can be successfully managed.

NURTURING PEOPLE

Previously, if you were loyal, did what you were told, minded your own business, and were generally pleasant, you could do well, and rise in the organizational hierarchy. Most people were asked not to think or innovate, but to do as they were told or to carefully interpret the senior managers' wishes. This concept of implementing the boss's ideas went all the way up the ranks to the highest levels of most organizations. Today, passive behavior and blind obedience are disavowed. Businesses need leaders who can coach, and employees who can think.

> Remmele Engineering Inc. is a modern contract machining, assembly, and automation equipment manufacturing company located in St. Paul, Minnesota. It is a company which values people. As one manager there put it, "You think education is expensive? Try ignorance."

Interprises reward entrepreneurship and risk taking, as opposed to blind obedience to executive orders. They seek only those skills which make them unique and world-class, and that keep customers in tight relationships with them. They do not try to hold all the skills and capabilities they need in-house. They ask employees not only to specialize, but to do so in a way that allows them to communicate with other employees, customers, and suppliers so that everyone can participate in improving work processes, and those of customers and suppliers. They recognize the inability to offer lifetime employment, and agree to work with employees to make them not only employable, but the best and most desirable employees throughout their working lives.

As the world changes more rapidly than before, skills just as rapidly become out of date. In the realm of computers, for instance, by the time you have become proficient at a system, it will have become outdated. The rapid obsolescence of skills and knowledge leads companies to make greater efforts to educate and train workers. In times past, they worried that the cost of education and training would be wasted if an employee left the company. Today, more and more companies are coming to the realization that not only is training necessary because of constant rapid change, but that if a person does leave—except in the case of intellectual property which is governed by law—their training would have soon become outdated in any event. It is also easier to manage a company if employees are up-to-date and employable by other companies, allowing a company to accept outsourcing work for other organizations in slack periods, and easing the ability of dismissed employees to find other jobs.

To this end, interprises work with employees to make them employable. They provide advice and opportunity, even career counseling and placement services. The employees, however, must be in charge of this process, as they learn to take control of their careers. They are expected to contribute ideas, behave in an entrepreneurial way, take appropriate risks, and seek and accept performance-based compensation. Pensions and healthcare plans that can be carried over when an employee leaves, and other portable benefits, foster a degree of employee independence, while stock and profit sharing are the incentives of a common destiny.

THE CORPORATE RESPONSIBILITY

Creating a functional, pleasant work environment can make a measurable difference to a company's goals.

> Lotte Bailyn is a member of a research team at Xerox Corporation, and a professor at MIT's Sloan School. In her studies, Bailyn found that work surroundings often led to problems of time management and hence to difficult pressures. A group of seventeen engineers were under pressure to get a new printer to market in eighteen months. Looking for time to concentrate, some had taken to coming in at 3 a.m. or staying until 10 p.m. To their surprise, the study revealed they spent 52 percent of their time meeting and interrupting each other. They then agreed to set aside quiet times in the morning when interruptions were banned. This experiment was so successful that the division head credited it for its first on-time project launch.

Paying attention to employees' deeper desires and needs may not sound like a very revolutionary idea, but companies can be shocked when doing something this simple produces good results.

> In the Dallas area, a group of forty sales and service people tried to work together to better serve their customers. They were so divided by culture and geography that they had a difficult time communicating, and each group blamed the other for problems. They had a common goal, however; the desire to reduce stress and improve their personal lives. It was this shared objective that eventually provided a focus for them to come together. It helped them gain a broader vision, and they started to exchange information and make joint calls. Sales exceeded projections and customer response times improved remarkably.

Jim Edwards, the controller at an administrative center, did a simple thing for his employees: He gave them a measure of control over their own schedules. Nearly half modified their hours, morale improved, and absenteeism dropped by 30 percent.

Paying personal attention to employees hardly sounds like a startling new idea, but generations of same-size-fits-all mass production organizations have led companies to regard people as cogs in a wheel, doing as they are told. The interprise depends on the knowledge of motivated people, who are empowered to define and solve problems. These are free spirits whose creativity and energy are harnessed by the interprise, instead of being held prisoner in a bureaucratic hierarchy. The new world of dynamic interaction gives people the freedom to follow their talents, even as it increases pressure by not guaranteeing the same job forever.

FROM LIFETIME EMPLOYMENT TO LIFETIME EMPLOYABILITY

Interprises like Texas Instruments and Motorola tell their employees that they cannot guarantee their job, but they can make them the most desirable employees wherever they are. The idea of the employee being coresponsible with the company for his or her destiny as well as that of the organization's is new. We are accustomed to placing sole responsibility for both in the hands of the company. Management has begun to understand and change and is now willing to work with the employees and seek their advice on behalf of the company. Employees at all levels are having a difficult time with the transition in relationship.

Prince is a Michigan-based, $800 million manufacturer of automobile interior and trim systems, with plants in the U.S., Holland, Mexico, and the U.K. In 1994 and 1995 they experienced 40 and 30 percent growth respectively. Each plant manager updates each shift weekly on all activities, and meets personally with each worker at least once a year.

They have an extensive employee training and education system, and encourage innovation by offering many imaginative awards, including the stolen base award for copying a good idea from someone else and implementing it. In 1995, they received 13,600 improvement suggestions from 2,700 people. They know and accept that skilled people may leave them for other companies, but that is no reason to reduce training. If an employee leaves to start his or her own business, they will often support that business, even adopt it as a supplier.

We are indeed facing a cultural change in the attitudes and responsibilities of leaders and employees. Trust and open, honest communication is a strange new opportunity for most organizations, as is genuine candor with respect to goals, finances, and other previously private information. Employees at all levels now need to partner with the organization they work with while they work there in order to move both the organization and their career along simultaneously. There are no impediments to operating this way except the old cultural assumptions and a fear of change. These are, of course, two large obstacles that may stand in the way of successful adaptation.

COMPANY-EMPLOYEE RELATIONSHIPS

During a recent workshop with human resource experts from over forty organizations, we explored the concept of offering employees something in exchange for the ability to offer lifetime employment. Following are the things we found that a company can provide to its employees:

1. Recognition and rewards
2. Portable benefits
3. Quality of life services and flexibility
4. Convenience services
5. Personal development
6. Transition assistance

7. Flexibility to individualized needs
8. Stock incentives
9. An enjoyable job

In a similar discussion of what an employee could offer his or her company, we found the following:

1. Career planning
2. Management of educational development
3. Maintenance of world-class skills profile
4. Management of health care, pension, and other benefits
5. Development and use of entrepreneurial skills to mutual benefit
6. Relationships both inside and outside the organization that increase personal and organizational effectiveness
7. Clear personal growth objectives with time frames
8. Organization or participation in group and self-improvement programs
9. Communication and presentation skills
10. Development of listening skills
11. Development of coaching skills
12. Development of leadership skills

For many participants, the surprise result in this exercise was not that such activities could be valuable for both employees and the company, but that the employee should take responsibility for his or her own development. The idea that the company would partner with the employee while they were employed, and provide transition help in finding the next job, was perceived as novel.

PRINCIPLES FOR INTERPRISE LEADERS TO OBSERVE

1. Invite employees and their spouses to discuss what they might value

2. Listen carefully to the economic security issues that concern them and try to find solutions
3. Be honest and open. Share real options and real constraints
4. Recognize your real and costly obligation to help employees take charge of their own destiny

The interprise encourages and attracts enlightened and empowered employees who are self-motivated, and take charge of their own destiny. They form a powerful ally to any company for whom they work, and they expect a stake or share in the income to which they contribute. They also seek recognition and a degree of independence that allows them to manage their own successful careers. These are the people weapon of the twenty-first century.

SUMMARY POINTS

⇨ The concept of people as assets is new and is just being recognized.

⇨ The old guarantee of lifetime employment is a thing of the past. The lack of control over the destiny of the people who work for the organization may initially be difficult for both management and employees to accept.

⇨ Employees need to be able to trust management. Management will have to respond to this need with open and honest communication about things that were previously considered confidential.

⇨ A company must let entrepreneurial employees know where their thinking and innovation are needed, and reward them appropriately when both succeed.

THINGS TO THINK ABOUT

☐ Does your company think of its employees as entrepreneurs?

☐ Do you take responsibility for your own employability?

☐ Does your company make a significant effort to constantly upgrade people's skills? Have you and your company agreed on a personal skills upgrade path?

☐ In the transition to self-management of their own careers, people will make mistakes. Is your company tolerant and helpful when this happens?

☐ Does your company improve employee morale in new and creative ways, or are they stuck behind old ways of thinking?

Chapter 21

LEADING
THE WAY

Leadership of an old-style company required order, and clear, visible systems, as well as reporting structures and precise job definitions. These were necessary for the product-focused company to work as reliably as an old-style Swiss watch. The interprise, on the other hand, succeeds because of the high degree of interaction with customers and suppliers, and within the organization. Successful interaction cannot be mandated and organized by management. You cannot order people to be motivated and successful entrepreneurs who find innovative business opportunities where others see only barriers.

To lead is to proact, not to react. Machiavelli said that a leader listens carefully, but heeds no unasked for advice. The leader sets the context, and in that context gives people freedom to act. The leader does not follow the consensus of the crowd, but proactively creates the circumstances within which stakeholders understand how much each should give, how much each can expect to take from the interprise, and how they can manage their conflicting requirements for the best mutual benefit.

Jan Carlsson, CEO of SAS, the Scandinavian Airline System, took his job in the late 1970s when the airline was losing $20 to $25 million per year. Six years later, a lack-luster airline was on its way to being a leader. Carlsson knew just where he wanted to go. He articulated his vision in a simple way: "I want to personally pick up the passenger in a luxurious limousine at their home, transport them, accompany the passenger to their hotel at the other end of the journey, go with them to the room without checking in, and find the suitcases waiting."

What better example is there of a corporate vision to become an interprise, part of the customer's business or lifestyle processes? He also saw how to accomplish that. "I calculate that SAS employees have 63,000 customer contacts every day. These are 'moments of truth.' If each such contact were pleasant, efficient, and successful, our airplanes will be full." He followed through on the vision, by changing the company. The old departmental structure was replaced by a self-managed team structure built around the routes the airline flies. In hiring, he emphasized the recruiting of extroverts who like to interact with people. Employees are systematically trained, with one week every year in the corporate leadership program. Anyone is empowered to solve any customer's problem if they can, so that a passenger need not be shunted from one specialist to another.

THE RESPONSIBILITIES OF LEADERSHIP

You often hear military metaphors in business: "lead," "win," "eliminate the enemy," and such, but the reality of military phrasing is often misunderstood. Leaders in business are like generals in the military, just as managers are like sergeants. The fundamentals of leadership are similar, whether in business or the military. Many people, however, have the mistaken impression, reinforced by generations of movies, that a general gets things done by barking orders and making sure that blind obedience drives his forces. Nothing could be further from the truth. The general cannot get anything done without the people who command troops in the field, and who must deal personally with the dead and wounded. When the general gives orders, he ensures that the orders are understood, that they take into account all reasonable facts, opinions, and objections, and that the aim and context are clear. He does not specify in exacting detail how a mission is to be accomplished; he sets goals and timetables, but leaves details to the field commanders. Business is no different.

The platoon sergeant has a different mission than the general, just as a line manager has a different mission than an executive. The sergeant is given an objective limited in time and in geography (you take the west side machine gun position on the next hill at 1700 hours today). When preparing and executing his mission, the sergeant cannot waste a single minute looking at the distance beyond his field of action. He needs to ensure that his people are correctly equipped, that they understand in minute detail what they are to do, and that in the confusion of combat they support one another rather than shooting each other by mistake. This is also the traditional role of the manager.

The leader sets the goal and context, the manager decides how to implement and execute. Both general and leader keep their eyes on the horizon and beyond; the manager and sergeant keep their eyes low, at the problems close by. The modern military force, and the modern interprise, have to be able to deal with unpredictable scenarios, and to do so victory depends not only on the perseverance of the soldiers or workers and the quality of the equipment, but also on the skillful deployment of knowledge. Both emphasize education, and make an effort to ensure that the goal of the organization is always clear to everyone. The sergeant's personality and example move the platoon, but the general's farsightedness and planning create the circumstances within which the sergeant must act, a relationship similar to that of the business leader and the manager.

An interprise leader worries about the goal and orientation of the business, market data and connections with customers, the quality of the supplier chain, the standard and availability of people to work in the organization, training, the organization of the corporation, and whether the people in the organization see the goal correctly.

LEADERS AND MANAGERS

The difference between leader and manager can be summed up in a phrase:

Leaders do the right things; managers do things right.

People work *with* a leader, but work *for* a manager. When working for a manager, most people feel tied down, unable to soar, delayed, and inhibited by needing the approval of the manager. Great leaders give people the feeling they are working with a contributing supporter. They feel the support and confidence of the leader. This generates a feeling of freedom and releases creative ideas. Many people have had the privilege of working with such a supportive leader. The interprise depends on supportive leaders and entrepreneurial workers, much more than the old-style company did. Leadership has replaced management as the driver of successful companies.

A leader needs the right mix of forcefulness and humility. He or she should be energetically ambitious, but not intimidating. For instance, at a business discussion between a supplier and its customer's president, the supplier's representative suddenly blurted, "This negotiation is unfair. You're more intelligent." The meeting reached an impasse caused by lack of trust. Had the customer's executive repressed his urge to show his cleverness, the negotiation would have concluded successfully, rather than leaving a residue of mistrust which was never fully overcome.

The leader cannot neglect to provide vision and support for the people who rely on him or her, nor can he or she afford to neglect the details of daily work. Action in the interprise must be focused in order for the interprise to be competitive. Without both vision and attention to detail, action is unfocused. The leader finds the right balance and a comfort level which suits both him or her and the people in the organization, between providing vision and support, and paying attention to detail. The leader develops the capacity not to underlead, nor to overmanage.

EVERY WORKER A LEADER

In every successful interprise, all employees need to be aware of the context in which the business works, so that when an unexpected situation develops, each employee is willing and trained to take responsibility, and can make a decision which is correct for the company. The alternative, which is to look for higher authority to solve the problem, or to give permission for an initiative, would be too slow to be useful.

The old-style company could succeed with a manager at the helm, and managers below him or her, each doing as they were told; the interprise requires a leader at the helm, and leaders throughout the organization. The interprise prospers only when the people of the interprise function as entrepreneurial, motivated leaders. It is not easy to lead leaders. In the old-style company, managers were sergeants; they pushed and cajoled people to get work out—the faster the better. In the interprise, managers act as generals. They motivate others. They set goals, give support, see that the resources are there when needed and help their people solve problems. They are supporters and coaches, not micromanagers. A modern manager takes great interest in the details of work, but sees his or her success as the success of others.

LEADERS COACH, MANAGERS COMMAND

The dynamic, interactive work which is part of the fabric of the interprise is not achievable when permissions are needed for every action. No customer or supplier wants to deal with someone who does not have authority to make decisions. Empowerment of people in teams is essential for an interprise. Leadership puts in place the people, work methods, and systems which make that possible. Management has moved from command and control to coach and coordinate. This holds for big and for small companies.

Iscar was started in a shack in 1952 by Stef Wertheimer, under conditions which all rational investors thought hopeless. Located in Galilee, they are today the fourth largest manufacturer of tools and drills in the world, with annual sales of $350 million. Wertheimer knows how to motivate and empower his people, who love him and constantly come up with new ideas. Chief engineer Harry Dickman, who together with Juval Bar-On installed a world-famous automated production line at their Carbide Division facility in Tefen says, "I cannot imagine another employer entrusting me with such a large budget, and then not bothering me, but just letting me do the best I can." That trust paid off. Not only does Iscar have excellent automation, but their expertise is now sought by other companies who hire them as automation consultants.

Legend has it that Bill Hewlett, who together with Dave Packard founded the Hewlett-Packard Corporation, knew how to nurture innovation. When someone would come to him with a bad idea, he would listen, then say, "I'll think about this, and call you tomorrow." The next day he would call and say, "You may want to think about the this-or-that aspect of the problem. When you have thought that through, come talk to me again." He was careful not to dampen the innovative spirit of his people, even when they had poor ideas.

In the old-style company, work was broken down into simple specialized tasks, each easy to understand. The company could be run by a manager whose outlook was limited to coordinating the efforts of specialized departments. Competitive advantage was attained by coordination and optimal allocation of resources between departments. Today's business environment is dominated by complexity, which cannot be managed by narrow specialization and detailed planning. Organizational complexity can be managed only by having initiative suffused throughout the organization. The job of the leader is then to express a clear goal, and make sure the people understand and can achieve it. This motivating leadership is more challenging than the formula-driven management style of the past.

THE LEADER'S CHALLENGE

Leadership of an entrepreneurial, adapting interprise in which people are continuously in contact with customers and suppliers, deciding themselves how to spend their time and making commitments in the name of the interprise, is an intense challenge for leadership.

> **The single most important leadership challenge is to clarify reality for the people in the organization.**

Leadership is the responsibility to define clear, achievable goals, in a timely fashion.

> The head of Mt. Sinai Medical School once said, "Most of my time I spend trying to tell people what is important. All the rest is commentary." To do that, he projects a clear, crisp vision, and a clear plan to implement it.

> Winston Churchill, the British Prime Minister during World War II, knew that people want their leader to define an inspiring goal, but they want it presented in simple terms. He made a special effort to phrase his memorable speeches in short sentences.

The goal must be clear to everyone throughout the organization. Leaders understand that need.

> The Israeli General Ethan sat in his command bunker to follow the progress of a squad sent secretly on a night mission to another country. A cryptic message came in: "Lost a soldier in the dark." He understood that the young lieutenant, trained to achieve his objective whatever the unexpected difficulties, would be torn between this objective and finding the soldier. He immediately spoke into the microphone, "Your mission has been changed. Your mission now is to find the soldier. I repeat, your old mission has been changed." Ten minutes later, the lieutenant radioed in that the soldier was found in the dark. The gen-

eral spoke into the microphone, "The mission is again to achieve the objective as before."

A concise picture of the competitive challenges of the interprise energizes and empowers the people of the interprise to pose the right questions and find solutions to problems.

In the late 1980s, the 3-M magnetic tape plant in Hutchinson, Minnesota, was faced with serious international competition. Over a period of four years, they reduced waste from 22 percent to 4 percent, reduced costs by 10 percent to 14 percent per year, while maintaining quality standards at under four defects per million. To achieve this, management minimized commands and lectures. They sent small teams of workers, armed with camcorders, around the world to competitors' plants. They saw for themselves the high standard of education, technology, and work ethic at their competitions' plants, and brought that picture back vividly on tape to their coworkers. The managers also arranged group bus tours for all the workers, to stores where their product was sold. They learned first-hand how they compared with their competitors. They heard about every detail, from the quality of the product to the convenience of the packaging, to the rapidity of on-time delivery. Management succeeded in having the workers understand the competitive context.

The biggest trap for an organization is that the collective vision of where they are and where they are going remains constant as competitive circumstances change. The efforts of individuals then become defocused and out of phase with reality in a changing environment.

Leaders know how to motivate people to higher levels of effort and success. They are tolerant of failure, provided it is positive failure, occurring when employees are striving towards significant goals.

The late Pete Estes, a former president of General Motors, was a leader's leader. He said, "If I want a job done well, I find a good leader who has failed once. He knows what it is to fail and will never fail again."

Leadership of the interprise is an exacting challenge, requiring a leader to succeed when juggling a seemingly impossible mix of aptitudes (see box).

To be a leader is to successfully juggle a mix of opposites. The leader:

- sees the general picture, but does not lose sight of details;
- deals well with people, but understands technical questions in detail;
- shows both controlled ambition and understanding humility;
- wants credit for success, and gives more credit to others than is due;
- is ruthless in pursuing business aims, and honest in acknowledging his or her errors;
- is affable and friendly in the crowd, yet not intimidated by the loneliness of leadership;
- exudes a sense of urgency, yet is patient with people doing their best;
- is an active thinker and a thoughtful activist;
- knows when to hold and when to fold, when to persevere and when to let go;
- is both protector and servant of his people;
- knows there is no substitute for experience, yet no time to gain sufficient experience.

> The leader manages the conflicting demands of the stake-holders to the interprise. These include:
>
> - the shareholders who usually want a high quick return;
> - the banks and creditors, who want to ensure that their loans will be reliably repaid;
> - the employees, who want high incomes and secure life-long jobs;
> - the suppliers, who want an understanding, high paying, reliable customer;
> - the customers, who rely on the interprise to support their business or lifestyle processes;
> - other partners to the interprise.

NEW ROLES AND NEW JOBS

The requirement that the interprise be adaptive as it maintains intense interactivity within itself leads to changing roles. In the old-style company, if you needed something bought, you went to a purchasing agent who specialized in buying. If you needed something sold, you went to a sales person who specialized in selling; designing was done by the design department, and so on.

> The legal counsel of Motorola, a successful interprise, said that if an issue comes to his office to resolve, then by the fact of its coming to him, he has failed right there. He sees his job as giving each team all the facts and access to all the expertise needed for the team to deal with legal issues. If an issue gets to him to deal with, it is a sign that he has not succeeded in his mission, which is to empower teams.

The change of organizational structure of the interprise, and the changing nature of leadership, have led to changed responsibilities of functional departments:

- a marketing department does not market; it sets up the system to support the operational team's ability to find and interact with customers;
- a purchasing department does not purchase; sets up the system for a team to purchase what they need;
- a design department does not design; it sets up the system of designing so that self-managed teams can design;
- a legal department does not negotiate contracts; it sets up the system for teams to negotiate contracts

The arm's length specialization that worked in companies a generation before, is not able to maintain sufficiently effective and speedy interaction in the interprise. The interprise succeeds when motivated people, working in teams, succeed.

Leadership was in the past a one-dimensional task, which could be performed reasonably well by even mediocre leaders. Leading the interprise is a multidimensional task, requiring an inspiring leader who can simultaneously manage many conflicting demands. In this multidimensional complexity, the leader cannot possibly get to hear of every business development, let alone make each major operational decision. Operational decisions must necessarily be left to empowered individuals in teams. Although this seems chaotic, the leader ensures that the interprise is in fact guided to success by empowered people in teams. Knowing how to empower people and trust them, while simultaneously understanding the limit of that trust, requires new standards of leadership. It is an opportunity for motivated leaders who would have been frustrated in the old hierarchic system.

SUMMARY POINTS

⇨ The interprise requires visionary leadership, not dictatorial bosses.

⇨ The leader's first job is to clarify reality, and set a clear "stretch" goal for his or her organization.

⇨ The new structure leads to new roles for old functions in the organization.

THINGS TO THINK ABOUT

☐ If you are a leader, have you provided your people with a clear, unambiguous, and personal message, so that they are aware their future depends on them, as part of the interprise, becoming interactive with customers?

☐ How many customer contact "moments of truth" are there in your business each day?

☐ When you lead, do people feel they are working *with* you or *for* you?

☐ Who are the stakeholders in your organization? What concerns do they have which you should address?

CHAPTER 22

CREATING THE
INTERPRISE

In the first part of this book, we showed how the new competitive environment is pushing businesses to do new things for the customer, and how to build long-term relationships rather than selling old-style products in new ways. In the second part, we saw that making money in the new economy requires tight, interactive working relationships with customers, suppliers, and partners. At the beginning of this third and final part, we looked inside the company to show that business success today requires an adaptive entrepreneurial environment with a very different leadership and organization than before. We have found that some fundamental business practices, such as the misleading and inaccurate cost accounting method, must be completely rethought so that companies can compete in the new business landscape. We have called the evolving, linked, dynamic, time-dependent business system agility. The interprise is the organization that successfully meets this challenge.

The brief overview in the three parts of this book demonstrate how the interprise is succeeding in the new, agile business environment of interactive processes, which is so radically different from the old arm's length business world. It is providing new and different opportunities for its customers, viewing each customer as an individual in a fragmented marketplace, and delivering a customer-specific, constantly-changing mix of product, information, upgrades, and service. The interprise competes by managing long-term relationships with customers, suppliers, and partners. It adapts its internal organization to suit the requirements of these relationships, rather than trying to force the relationships to fit the structure of the company. The new corporate structure requires teaming, and emphasizes the value of people and knowledge, guided and coached by a visionary leader. The interprise measures itself by entirely new metrics in every aspect.

The reader who is interested in no more than an overall picture should stop reading at this point. What follows is intended for people and organizations who want to formulate an action plan for moving from arm's length business to the interactive, agile, and competitive interprise. It is intended for upper and middle management, and the consultants who serve them, and allows you to select the parts of the organization you want to pay attention to, in order to prioritize your efforts in the move towards becoming an interprise. The methods and worksheets are derived from the experience of the Agility Forum at the Iacocca Institute of Lehigh University, and are intended for use in a facilitated group setting where all constituencies of the organization are represented.

THE METHOD

Our method in these last chapters is based on a generic model of an interprise. It consists of a series of worksheets filled out by consensus and after discussion, and ends up with a prioritized list of issues to be dealt with in a company. The model identifies the distinctive kinds of market forces driving business change today, the attributes of an enterprise capable of thriving in that market environment, and the organizational infrastructure required to support those attributes. While not claiming to be either unique or exhaustive, the model reflects the views of at least a thousand people—from CEOs to operational personnel—at hundreds of companies of all sizes across all sectors of the economy. The model captures the significant broad-brush thinking of U.S. companies today. It makes an excellent checklist, therefore, for any company's calculated response to its competitive situation.

Every company is different, and comes with its own set of circumstances. There is no single migration plan to becoming an interprise. The plan is necessarily context dependent. No company can, or needs to, reform every business system. The approach outlined here allows a company to characterize its own competitive context, and to make the appropriate context-specific plan for itself.

GETTING STARTED

Before deciding where to start with the reorganization involved in becoming an interprise, you need to decide who should take the initiative for that move. Although change must eventually be supported and driven from top management, there is no way to know where islands of change will develop. They will develop and coalesce into a new way of doing business. Management must be open to this.

It is necessary to see a picture of the interprise as a whole, in order to migrate from the old corporate structure. We have dealt with the subjects one at a time, but these separate pieces are facets of a single picture. In the mass production mind set, functions could be broken up into separate specialties with minimal interaction between them. This is simpler, but in today's competitive environment it is inefficient. Today's successful businesses succeed by managing the complexity of a large linked system. Businesses cannot afford to maintain barriers within the organization as before, separating functions such as finance, purchasing, sales, design, or production, nor can there be barriers outside the organization between suppliers, the company, and its customers. Naturally, an organization cannot be a chaotic collection of uncontrolled entrepreneurs, each using whatever people and resources their enthusiasm calls for. Boundaries and points of responsibility are still needed, but today these are porous and temporary divisions that are constantly relocated as opportunities come and go. Responsibility requires seeing the good of the whole organization rather than of just one narrow division.

The following table compares and contrasts the characteristics of an agile interprise versus an old-style enterprise:

The Agile Interprise	*The Old-Style Enterprise*
• Sees change as a competitive opportunity	• Sees change as a problem to be avoided
• Has interactive relationships with customers and suppliers	• Has arm's length relationships with customers and suppliers
• Sells a customer-individualized, varying mix of product, service, and information, or uses product as a platform	• Sells a fixed menu of products or services with service or the product as the goal
• Has an adaptive organizational structure	• Has a rigid organizational structure
• Empowers cross-functional workers	• Encourages narrowly specialized workers who will not do another person's job
• Leads by coaching	• Leads by dictating
• Is powered by a continuous input of new knowledge	• Has no significant input of new knowledge
• Uses business systems measured by time-dependent rate metrics	• Uses business systems measured by static metrics which take no account of time rate changes

GENERATING AN ACTION PLAN

There are four questions that are answered by any company seeking to transform itself into a more powerful competitor by exploiting the interprise model.

1) What are the relevant market drivers of change to which my company is responding—either reactively or proactively—by becoming an interprise?
2) What are the attributes that my company should possess in order to assimilate those drivers into its operations?
3) What new capabilities should my company acquire if it is to possess those attributes?
4) What business process changes will be necessary to support the new capabilities?

We have used these questions as the basis of our Generic Model below.

THE GENERIC MODEL

I. Market Forces Driving Business Change

1. Intensifying Competition
2. Fragmentation of Mass Markets
3. Cooperative Production Relationships
4. Evolving Customer Expectations

II. Enterprise-Level Attributes

1. Solutions Provider
2. Collaborative Operations
3. Adaptive Organization
4. Knowledge-Driven Enterprise

III. Enabling Infrastructure

1. Interoperability
2. Reconfigurability
3. Flexibility

IV. Business Processes

1. Demand Identification or Creation
2. Product and Service Realization
3. Demand Fulfillment
4. Enterprise Management
5. Metrics

In the Generic Model, action at each level requires support from the level below, as represented in the following chart.

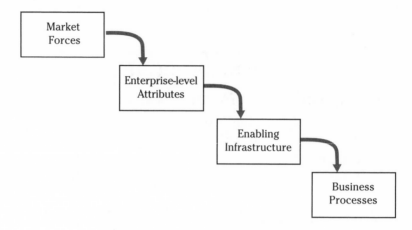

Figure 22.1 The Generic Model.

We will use the Generic Model as an assessment tool by asking you to select and prioritize those market drivers, enterprise attributes, and enablers that you judge to be most significant for your company. These selections then become the basis for a coordinated, strategic, and integrated approach to rational business process reorganization, adapted to the competitive situation of your company at this moment in its commercial life. The remaining chapters of the book follow the flow of this model, from market drivers to business process change, pulling together the ideas we have discussed in earlier chapters, and identifying their applicability to your company.

FILLING OUT THE WORKSHEETS

To assist you in this, blank worksheets in the form of matrix charts are provided for each of the four interprise model elements. A complete set of blank worksheets is also provided at the back of the book for your convenience. We preface each worksheet with several questions intended to stimulate thought on the value of the different threats and opportunities to your company in light of the interprise model. The group facilitator may find these questions helpful in generating discussion.

At the end of chapter 25, a master worksheet will capture your highest ranking selections, and identify their impact on your plans for business process change. For comparison, two sample sets of filled-in assessment worksheets—one for a service company and one for a manufacturing company—are included, along with a brief description of how they came to make their selections. These worksheets come from Agility Forum assessment projects at real companies, whose names are disguised here for obvious reasons.

How agility and the interprise model are used for competitive advantage will always depend on context. The worksheets and their wording represent one implementation of the subject matter of this book as developed by the Agility Forum for the organizations they deal with. Where the methodology and words are adapted to suit different companies and needs, other implementations are always possible.

THE MARKET FORCES DRIVING BUSINESS CHANGE

The first step, and the heart of any business case for transforming an enterprise into an interprise, is to identify the market forces to which creating an interprise is a response.

Intensifying Competition

a) While the "clockspeed"—to use the words of MIT Professor Charles Fine—is very different for different industries; every company's clockspeed is running faster today than it did fifteen years ago.

Facilitator's Warm-Up Questions

◇ What is the effect of accelerating change in your markets? Is it a driver of change in your company's operations?

◇ If yes, what schedule do you need to follow for rollout of new products and/or service introductions?

◇ Does the drive come from a high rate of technological innovation in your industry, or from competitors who, through organizational innovations, are able to bring new products to market faster than you can?

◇ Would bringing products to market more quickly improve

the competitiveness of your company? How much more quickly would that have to be? Is the drive for rapidity coming from your customers, your competitors, or is it a strategic choice you can make in order to create competitive advantage?

b) With the intensification of competition comes constant pressure to lower costs and increase productivity. Competition has become more intense in recent years through the decreasing cost of information and increasingly capable communication technologies. One consequence is the globalization of competition as companies seek out the lowest cost providers of capabilities wherever they may be located.

Facilitator's Warm-Up Questions

◇ To what extent is your competitor's exploitation of information, or of communications technologies, a driver of business change for your company?

◇ How effectively are you exploiting information as a product or service? How effectively are you using information as a basis for new relationships with suppliers, customers, and partners?

◇ How well are you using available communications technologies to reduce costs, improve productivity, or exploit global resources?

◇ Do you have a proactive strategy in place to get ahead of constant pressure to reduce costs and increase productivity?

Fragmentation of Mass Markets

a) The fragmentation of mass markets into niche markets has been made possible by the availability of technology. The entry costs for competing in niche markets, such as publishing, advertising, engineering design services, precision machining, audio and video studio services, software development, product prototyping, financial services, and contract manufacturing, continue to decline, creating new kinds of competitors for large companies that are scaling mass market-based operations down in order to compete concurrently in multiple niche markets.

Facilitator's Warm-Up Questions

◇ Has your company responded in a systematic and coordinated way to the challenge of targeting niche markets? How have your planning and marketing processes changed to support the shift from mass to niche markets?

◇ Are you vulnerable to competition from competitors who are exploiting the declining cost of relevant technologies to target a few of your market niches, even the most lucrative? Are you exploiting lower cost, modular technologies to compete in lucrative niche markets that you did not previously compete in?

b) Product market lifetimes—especially for information and service products—have become shorter than ever before, with frequent model changes even within short lifetimes. As a result, the profit window of a successful new product is likely to be very narrow as imitative competitors rush in. To sustain profitability, new windows must be opened by introducing still newer models and newer products, as Hewlett-Packard, for example, has done successfully with generation after generation of ink jet and laser printers, or U.S. Robotics with modems.

Facilitator's Warm-Up Questions

◇ Do you have a product development strategy adapted to such a dynamic marketplace? For example, do you require that new product or service designs support an evolutionary development strategy, allowing them to serve as platforms for multiple model changes and long-lived product families? Do your sales and marketing policies reflect this strategy? What about your supplier relationships and logistics processes?

Cooperative Production Relationships

a) Throughout the business economy, companies are finding that the cost and time involved in arm's length relationships with customers and suppliers has led to very uncompetitive capabilities, driving a change to cooperation.

Facilitator's Warm-Up Questions

◇ Is your competition exploiting cooperative relationships with suppliers, customers, and partners more effectively than you are? Do you have mechanisms in place for continuously reviewing make-buy trade-off decisions? Are your competitors exploiting global sourcing, marketing, or distribution channels? Are you?

◇ Are you receptive to initiatives by your suppliers, customers, and partners to change your processes in ways that they suggest, to your mutual benefit?

b) Changes in labor-management relationships within companies are also a driver of business change today. So is the rapidity with which a company's resource constraints can change. On a time scale of months, and sometimes weeks, a company's growth, or its ability to exploit a market opportunity, can be constrained by available capital, by labor or knowledge, by sales channels, or by supplier capacity, capability, or lead times.

Facilitator's Warm-Up Questions

◇ How well are you managing dialog about the potential benefit of significantly different labor-management relationships compared with the ones now in place in your company? Could you imagine a detailed enough restructuring of those relationships which would make a convincing case to both sides? How clear is the case that these relationships hinder your current competitiveness? How are your competitors changing workforce relationships to gain a competitive advantage?

◇   Is there a single, constant constraint to your company's competitiveness or ability to exploit new market opportunities? If in recent years there have been multiple constraints of changing weight, is the company tracking these changes in order to anticipate future requirements for overcoming them?

Evolving Customer Expectations

Both commercial and consumer customers are increasingly demanding life-cycle product support in place of limited lifetime performance warranties. Expectations of rapid delivery of orders and of rapid time-to-market with promised new products make for an extraordinarily demanding marketplace by traditional, mass production-based, business standards. Quality continues to be a driver of business change. The definition of quality has evolved from reliability, which won market share in the 1970s, but no longer does, to customer satisfaction in the 1980s, and now to exceeding customer expectations, "delighting" customers.

Facilitator's Warm-Up Questions

◇   A growing number of companies use the physical product they manufacture as a platform for the information and services which bring in the profit. Is your company still selling products or has it shifted the center of value it adds for its customers to information and services? Does the support that you offer customers in your marketing and after sales focus on product use, such as reliability, or does it focus on customer needs, such as information and service added to products? Are your competitors redefining the nature of customer support ahead of you? Is enhanced customer support an important pre-sales competitive advantage for you?

◇   Is the desire to cost-effectively individualize products and services driving business process change in your company? Is it for your competitors? How valuable would more rapid time-to-market with new services and products be? How well do you compare with your competitors on this? What would be the advantage of adopting a strategy of bringing new services and products to market much faster than you do today?

◇ How is quality defined in your company at the operational level? Are you still measuring quality as reliability only? How well do you know the market value of quality for your products and services? How valuable is reliability to your customers? How valuable would exceeding their expectations be?

Instructions for worksheet 1: For each driver of business change listed, assign a value of 1 (low), 3 (medium), or 5 (high) in the first column, reflecting the importance of that driver in your industry or markets. In the second column, assign a value of 1, 3, or 5 to the threat posed by that driver to your company. For example, a driver could be very important to your industry but poses no threat to your company. Calculate the value of that driver to your company by multiplying the product of the importance and the threat and entering that value into the third column.

Worksheet 1   **Market Forces Assessment**

Date _____ Company Name _____

DRIVERS OF BUSINESS CHANGE	IMPORTANCE in your Markets	X THREAT to your Company	= VALUE to your Company
INTENSIFYING COMPETITION			
Rapidly Changing Markets			
Declining Cost of Information			
Improving Communication Technologies			
Pressure on Costs/Productivity			
High Rate of Innovation			
Decreasing Time-to-Market			
Global Competitive Pressures			
FRAGMENTATION OF MASS MARKETS			
Growth of Niche Markets			
High Rate of Model Change			
Shrinking Product Lifetimes			
Shrinking Profit "Windows"			
Low Niche Market Entry Costs			
COOPERATIVE BUSINESS RELATIONSHIPS			
Inter-Enterprise Cooperation			
Interactive Business Relationships			
Increasing Outsourcing			
Global Sourcing/Marketing/Distribution			
Shifting Resource Constraints			
Labor-Management Relationships			
EVOLVING CUSTOMER EXPECTATIONS			
Individualized Products/Services			
Life Cycle Product Support			
Rapid Time-to-Market			
Rapid Delivery			
Changing Quality Expectations			
Value of Information/Services			

Scale
1 = Low Importance/Threat
3 = Medium Importance/Threat
5 = High Importance/Threat

Reproduced by permission of The Agility Forum
125 Goodman Drive, Bethlehem PA 18015
Tel (610) 758-5510   Fax (610) 694-0542

CHAPTER 23

MAKING IT
WORK

In this chapter, we will take a look at the new market opportunities which can derive from changing an enterprise to an interprise.

The worksheets in this chapter allow you to keep track of the value to your company of the attributes in the model associated with selling solutions and engaging in collaborative operations and identify the distinctive enabling capabilities that will be required to support those attributes. Your selections of the attributes that you feel are most important for your company to acquire should reflect the selections made at the end of chapter 22 as the most significant drivers of change for your company.

More and more companies are looking beyond the hardware product or specific service they provide, to being part of their customers' processes—work processes for business customers, lifestyle processes for consumers. What sells Ross Operating Valves to its customers is the opportunity to codesign customized pneumatic valves and receive a prototype within days without paying a premium. What sells Harley-Davidson motorcycles to its customers is image: a particular look, feel, and subjective gratification that is only loosely related to product quality or performance. The continuing growth of the electronics content in automobiles, which sup-

251

port operation of the vehicle as well as anti-theft devices, enter-
tainment options, and communication and navigation services, is
an example of a product serving as a platform for information and
services. Indeed, over 90 percent of the profits earned by U.S. auto
dealers come from parts and service; only 5 to 10 percent come
from car sales. The greatest profit center for Harley-Davidson dealers is the line of insignia clothing they carry!

Information and services are often far more profitable, than physical products, and they can be individualized into a customer's
processes. Physical products become more valuable in a solutions-
driven marketplace to the extent that they are designed to support
customization and/or information and service packages.

ENTERPRISE-LEVEL ATTRIBUTES

SOLUTIONS PROVIDER

A powerful interprise works with its customers, and is sought by
new customers, because of its ability to support their processes.
The interprise emphasizes individualizable, information-rich products and services, with a high introduction rate, and frequent model
changes within each of its product families and individualized delivery and/or production services. This is achievable even with
high volume, as shown by Nike sports shoes, John Deere seeders,
and Toshiba portable computers.

Facilitator's Warm-Up Questions

◇ Do you have a strategy in place for identifying niche market opportunities? Are you creating new market niches
   by creating variations of your products and services?
   Would individualizability enhance the value of your products and services? Are your current products easily individualizable, upgradeable, and reconfigurable to support
   changing customer requirements over time?

◇ Into how many models do you segment your product families and how rapidly do those models change? Do you
   use information and services to extend your product
   portfolio and add value for market segments?

◇ Are you still producing to forecast? What is the cost incurred by the mismatch between forecasts and market reality? How valuable would it be if you produced to order only? How does your current new product concept-to-cash time compare with the time it took five years ago? How valuable would it be if you could introduce new products twice as quickly?

◇ Do you know enough about your customers to show them how to be better customers for you? Do you know enough about your current customers to be able to create with them products and services that they do not yet know that they need? Do you know why companies that you feel should be your customers are not? Do you know what it would take to convert the most valuable of them into customers?

Instructions for worksheets 2 to 5: For each agile enterprise attribute, enter a value of 1 (low), 3 (medium), or 5 (high) in the first column, reflecting your assessment of the importance of that attribute in your industry. In the second column, enter your assessment of the competitive opportunity that that attribute offers your company. For example, the importance might be high, but the opportunity low because your company already possesses that attribute. Calculate the value of each attribute for your company by entering the product of importance and opportunity in column three.

Worksheet 2   **Agility Attributes I:** *Solutions Provider*

Date ——————— Company Name ————————————————

SOLUTIONS PROVIDER ATTRIBUTE	IMPORTANCE X in your Markets	OPPORTUNITY = for your Company	VALUE to your Company
Niche Marketer: High Product Diversity			
High New Product Introduction Rate			
Frequent Model Changes			
Rapid Concept-to-Cash			
Cost-Effective Low-Volume Producer			
Production to Order			
Individualizable Products/Services			
Life Cycle Design Methodology			
Open Architecture Product Design Philosophy			
Life Cycle Product/Customer Support			
Information/Services-Rich Products			
Enduring, Proactive Customer Relationships			
Proactive Marketplace Change Agent			
Solution-Based Marketing Policies			
Extraordinary Quality Standards			
Market-Opportunity Pulled Production			

Reproduced by permission of The Agility Forum
125 Goodman Drive, Bethlehem PA 18015
Tel (610) 758-5510   Fax (610) 694-0542

Scale
1 = Low Importance/Opportunity
3 = Medium Importance/Opportunity
5 = High Importance/Opportunity

COLLABORATIVE OPERATIONS

For the interprise committed to selling solutions, collaboration across functions and divisions within a company as well as with other companies is the operational strategy of first choice. When a problem needs to be solved, the first question is "Who can solve this problem?" If that turns out to be someone from another department, or from another division of the company, or from a supplier or a customer with a stake in the problem and its solution, the problem requires creating a team of these people. Creating such teams becomes the rule rather than the exception. Trying first to operate in a self-contained way and turning to others only as a last resort, is a luxury that companies cannot afford when dealing with dynamic customer opportunities.

Facilitator's Warm-Up Questions

◇ Does your company have systems in place that support a collaborative work environment? Is it easy to create teams of people with appropriate knowledge and skills to solve important problems, regardless of their departmental or divisional affiliation? How easy is it to add outside people to the team? How quickly could someone in your company identify people in other departments with competencies that would help them? How quickly could you identify an outside competency you needed access to? Do you have reward mechanisms in place that reinforce cooperative behavior?

◇ Are cross-functional groups that form in your company able to work together without being physically near each other? Are your customers and your suppliers involved in new product design and development processes?

◇ Would it be relatively easy or difficult for your company to enter into a virtual organization partnership? Would your accounting and cost management methods be obstacles? Would the legal department be a help or a hindrance? Do you have clear enough intellectual property rights and information sharing policies to participate in such an intimate joint venture? Do you recognize market opportunities that you could profit from by using the virtual organization model, but that you are not now exploiting? Do you have mechanisms in place that would support making a business case for entering into a virtual organization partnership?

◇   How widely does your company currently employ electronic commerce? How deeply does your ability to communicate electronically go down into your supplier base and up into your customers? How integrated is the flow of information across the various departments and divisions of your company?

Agility Attributes II: Collaborative Operations.
See instructions for worksheet 2, on page 253.

Worksheet 3   **Agility Attributes II:** *Collaborative Operations*

Date ——————— Company Name ——————————————————

COLLABORATIVE OPERATIONS	IMPORTANCE in your Markets	X OPPORTUNITY for your Company	= VALUE to your Company
Cooperation = Operations Strategy of First Choice			
Concurrent Operations			
Integrated Product and Process Development			
Integrated Comprehensive Enterprise Processes			
Interactive Customer/Supplier Relationships			
Virtual Organization Partnering			
Electronic Commerce Operability			
Proactive Information Sharing Policies			

Scale
1 = Low Importance/Opportunity
3 = Medium Importance/Opportunity
5 = High Importance/Opportunity

ENABLING INFRASTRUCTURE

The particular technologies, software, and organizational infrastructure that an interprise will need to put in place in order to operate collaboratively, and to produce solutions instead of products,
will depend on a company's industry, markets, and size. There are,
nevertheless, three general characteristics of the infrastructure of
any company responsive to the market forces described in chapter
22. They are the hallmarks of a company committed to the agile
business paradigm and to the interprise organizational model:

- Interoperability (Operational Integration)
- Reconfigurability
- Flexibility

When working with the interprise model to identify the attributes
that you believe your company needs to acquire, you must assess
the need to build interoperability, reconfigurability, and flexibility
into the implementation of those attributes. Although this discussion will not be encapsulated on a worksheet, it is nevertheless important to bear in mind these infrastructure needs when
formulating an action plan such as the one we are leading up to in
chapter 25.

INTEROPERABILITY

Almost every company has legacy problems in the form of existing
heterogeneous systems, processes, and procedures that have ac-
cumulated over a period of years. The term "legacy problem" was
originally applied to incompatibility problems associated with computer hardware and software, such as incompatible equipment or
programs. Different companies, and different groups within the
same company or even the same department, are all too often unable to share data or create common operations because of these
problems. Past decisions represent too great an investment to
abandon, yet their perpetuation impedes the functional integration
of operations within a company and interactive operations among
companies.

There are, however, legacy problems in companies other than computer-related ones. There are organizational and managerial legacy problems—inherited values and ways of doing things, control structures, and reward mechanisms—that can pose greater obstacles to intra and interenterprise cooperation than incompatible hardware and software.

Even though the virtues of cooperation loom large, it is not feasible, for a variety of technical and financial reasons, to insist on uniformity of equipment, software, and processes as the preferred means of achieving the integration necessary to support cooperation. The goal is functional integration, and that is achieved through interoperability. The Internet is a powerful example of how communication can be established, and digital information shared, independent of hardware and software. The requirements are appropriate design and strict observance of communication and information exchange protocols or standards.

Interoperable systems—information systems, production processes, and business practices—enable physically dispersed teams to work together. They enable intensive interactive working relationships with suppliers and customers. Once achieved, interoperability makes comprehensive, distributed, and timely access to information across a company's entire value chain possible. That, in turn, allows the creation of real-time management tools and enterprise-level processes and metrics.

Facilitator's Warm-Up Questions

◇ How interoperable are the systems and processes currently in place in your company? To what extent is existing interoperability being exploited to reduce costs and create new capabilities?

◇ To what extent is a lack of interoperability an obstacle to creating enterprisewide processes? To what extent is it an obstacle to creating value-adding interactive relationships with suppliers, partners, and customers? How valuable would interoperability be for your company, internally and externally? What are the costs of not having full interoperability?

◇ What are the barriers to implementing interoperability
and how might they be overcome? How much of an effort
is your company making to identify the standards it
ought to support and the strategy it ought to follow in order to achieve interoperability? By what date should you
aim to reach interoperability of 50 percent of your operational processes?

RECONFIGURABILITY

Configurability throughout the enterprise leads to cost-effective individualizability and solutions-based marketing based on products
and services that are easily upgradeable and reconfigurable. It also
results in a company's adaptability to unpredicted change. Reconfigurable communication and information systems, equipment, production and business processes, as well as reusable, modular, and
scalable equipment, processes, designs, and software, are all powerful enablers of an interprise. Reconfigurability is a primary deter-
minant of the speed with which new customer opportunities can be
exploited, and of the cost of that process.

Facilitator's Warm-Up Questions

◇ Is reconfigurability currently characteristic of your company's systems, processes, and practices? How valuable
would widespread reconfigurability be if it could improve
your company's proficiency in adapting to marketplace
change and recovering from "crashes"? How much cost
to your company is due to duplication caused by lack of
reconfigurability?

FLEXIBILITY

Flexibility complements interoperability and reconfigurability by
bringing the human dimension into the equation. Flexible organization structures enable a company's operations to be adapted to
changing market opportunities, making resources available in ways
that will have the greatest impact on the company's bottom line.

The flexible organization of work enables the routine creation of properly supported, opportunity-pulled teams. These are made up of people with the competencies required to do the best job, no matter where those people are located, or whether they are part of the company, its suppliers, customers, or partners. Flexibility is further discussed in detail in chapter 24 under the heading Adaptive Organization.

AGILITY ATTRIBUTE ENABLERS

Interoperability
Operational Processes and Systems
Business Practices
Information Systems
Integration Supporting Standards
Real-Time Management Tools
Comprehensive, Distributed Information Access
User-Driven Interfaces
Support for Physically Dispersed Teams

Reconfigurability
Modular and Scalable Processes and Systems
Integrated Product and Process Design Tools and Procedures
Rapid Product Development Tools
Reconfigurable Operational Technologies and Processes
Reconfigurable Information Systems
Upgradeable and Reconfigurable Products and Services

Flexibility
Work Organization
    Optimized, Proactive, Cross-Functional Teams
    Proactive Learning Programs
    Competency-Based Compensation Programs
    Social Values Internalization Programs
Management Philosophy
Organizational Structures

CHAPTER 24

MASTERING CHANGE

I n chapter 22, we collected the new drivers of business change under the headings Intensifying Competition, Fragmenting Mass Markets, Cooperative Production Relationships, and Evolving Customer Expectations. In chapter 23, we focused on the distinctive outputs of the interprise: providing solutions to customers' problems rather than selling them products; and on its distinctive inputs: cooperative working relationships across the company as well as with suppliers, customers, and partners. In this chapter, we will focus on the distinctive internal features of the interprise, its adaptive organizational structure, and the value it places on people, information, and knowledge.

ENTERPRISE-LEVEL ATTRIBUTES

Adaptive Organization

Today, a company has to sell individualizable, information- and service-enhanced products to demanding customers in rapidly changing niche markets against aggressive global competitors. That kind of environment discourages managing from the traditional perspective and encourages organizing internal operations to support external relationships. This will determine a company's ability to exploit, preferably proactively, new market opportunities with narrow profit time windows.

Throughout the book we have suggested organizational changes which facilitate a company's move from selling product to arm's length customers to becoming an interprise, and an interactive part of its customers' processes. We now group these suggestions into three categories.

a) Adaptive Management

Command-and-control management presumes that the most important relationships for a company are the internal ones—the ones that managers can have control over and whose functioning they can command. The result is a company with centralized authority, fixed lines of responsibility, and hierarchically organized decision-making. In such a company, initiative and accountability for the success of the company decrease as you move "down" from the executive suite to the workplace. Adaptive management, on the other hand, moves away from hierarchy, to a hypertext type of organization where a person or team can access any other person or resource as needed without seeking time-consuming permissions.

Facilitator's Warm-Up Questions

◇   Is your company managed to fit the needs of external relationships, or do you try to make external relationships fit your company structure? Would your colleagues recognize the difference between these two modes of management with only a brief explanation on your part? Is there a recognition in your company of the new value of external relationships?

◇   Are the executives in your company motivational leaders? To what extent do managers set goals and proactively provide empowering support systems for those reporting to them? Do managers work to create a culture in which personnel react to change as new opportunities rather than as threats? Is the ability to motivate one of the skills that is consistently evaluated, supported by training and rewarded in your company?

b) Adaptive People

Once a company's external relationships become drivers of the organization and of management of its internal operations, command-and-control becomes dysfunctional, as do centralized authority, fixed lines of responsibility, and hierarchically organized decision making. In place of command-and-control, the role of management becomes one of motivation and support, inspiring initiative and accountability from all levels of the workforce, and supporting all levels with the changing resources they need for the company to thrive in a networked mode of operation.

Facilitator's Warm-Up Questions

◇ Are entrepreneurial personalities, regardless of their job category, recognized as an asset or a nuisance in your company? Are there reward and support processes in place for such people? Are education and training programs structured to reinforce entrepreneurial behavior?

◇ To what extent is decision making distributed away from the executive center of your company? Is an effort being made to move decisions to the operational level? Are operational personnel being provided with training to enable them to make decisions that were once made for them?

◇ Are mechanisms in place, or being developed, to coordinate distributed decision-making without rigid centralization? Are personnel provided access to, and authorized to share with others, including appropriate people in other companies, timely, comprehensive, and accurate information about your operations?

c) Adapting to the Customer

Distributed authority, decentralized decision making, and an adaptive organization of work become necessary to enable a company to become part of a customer's processes, and follow the customer wherever he or she leads. This implies a shift to lines of responsibility that change as a function of the internal and external relationships that have to be supported. It also means reconfigurable physical facilities and equipment, flexible production processes and business practices, and an adaptive information system. The goal of management now becomes the creation of a timely, customer opportunity-driven organization that is proficient at changing its processes rapidly, robustly, and cost-effectively enough to create and exploit new market opportunities.

Facilitator's Warm-Up Questions

◇ How valuable would it be for your company to implement a customer-interactive organization? How well does your company's ability to change internal processes and external relationships match the rate of change of customer opportunities? How well does your ability to develop new products and services, modify existing ones, and change the mix of the products and services you now offer match the rate of change in your markets? How flexible is the organization chart in your company? Does it dictate the distribution of people and their skills, or does the need for people with particular skills drive the organization chart?

◇ Does your information system support the demands that people are placing on it, or would place on it if you adopted this mode of operation? Are you hampered by software "legacy" problems? Are efforts being made to solve them? Is the lack of information integration within your company and with suppliers and customers perceived as a problem and an obstacle to growth?

Agility Attributes III: Adaptive Organization.
See instructions for worksheet 2 on page 253.

Worksheet 4   Agility Attributes III: *Adaptive Organization*

Date _____ Company Name _____

ADAPTIVE ORGANIZATION	IMPORTANCE in your Markets	X OPPORTUNITY for your Company	= VALUE to your Company
Motivational Management Philosophy			
Coordinated, Decentralized Decision-Making			
Adaptive and Reconfigurable Physical Resources/Processes			
Adaptive Work Organization			
Timely, Opportunity-Driven Organization			
Change Proficient Organization			
Adaptive Information System			
Distributed Business/Production Processes			

Scale
1 = Low Importance/Opportunity
3 = Medium Importance/Opportunity
5 = High Importance/Opportunity

ENTERPRISE-LEVEL ATTRIBUTES

Knowledge-Driven Enterprise

Continuously learning people, information, and knowledge are becoming the key differentiators between successful and unsuccessful companies. Successful companies are the ones that enable people to learn and use information and knowledge to drive the development of a continuous stream of high customer value services and products. Furthermore, it is increasingly the case that the people that need to use information and knowledge to a company's competitive advantage are not only the employees of that company. As we have emphasized throughout this book, it is the flow of information and knowledge across the boundaries that previously separated a company from its suppliers, customers, and partners, which has special value.

A number of consequences follow from a commitment to the goal of enabling people to learn and use information and knowledge creatively.

a)   Knowledge Assets

A company is organized and managed in a way that motivates its workforce to be innovative across the total spectrum of processes associated with the production, distribution, and support of products and services. Open communication and information sharing policies across the workforce are powerful determinants of success, and cross-functional training emerges as an obviously worthwhile investment.

Facilitator's Warm-Up Questions

◇   Are the knowledge, skills, and initiative of your personnel perceived by your customers as your company's key competitive assets? Should they be? Are they perceived as key assets internally? Is this perception reflected in existing reward mechanisms? Would appreciating the roles played by people and knowledge in your routine operations attract new customers to your company?

◇   Is there an active flow of information and sharing of knowledge with your suppliers, customers, and partners? Is the concept of a learning interprise recognized as the next step in the evolution of a learning enterprise?

b) Knowledge Renewal

A company's organization must reflect the innate dynamism of the information and knowledge applicable to that company's market opportunities. The half-life of knowledge, and of the value of any particular piece of information, keeps shrinking. The attributes of an Adaptive Organization, as discussed above, enable the fluid reconfiguration of physical and human resources. New, enterprise-level performance metrics, in place of local operational performance metrics, are necessary to reinforce and correctly evaluate the effectiveness of information- and knowledge-driven operations. It is continuous education and training, however, that provides the armature around which to build a continuously learning organization. The objective is a company whose operations are centered on internal and/or networked expertise, rather than a company whose operations are centered on procedures.

Facilitator's Warm-Up Questions

◇   Is multiskill, cross-functional training routinely provided to personnel? Is it required as a matter of policy, in order to cultivate a learning organization? Is your company committed to continuous education and training? Is it treated as a cost or as an investment?

◇   Are education and training limited to current job performance or are they open-ended processes, to encourage learning regardless of immediate applicability? Are personnel supported by just-in-time access to immediately relevant information and skills acquisition, in addition to traditional classroom learning programs?

c) Knowledge Inventory

Management should know what its employees are capable of knowing, what their potential for learning is, and how that potential needs to be expanded through recruitment. In addition to regular "competency inventories," a company needs to implement processes that capture the often tacit and inevitably evolving knowledge that its personnel apply to the performance of their jobs. This is especially important when personnel, in effect, continuously reinvent their jobs in the course of continuously improving processes.

Facilitator's Warm-Up Questions

◇ Does the management of your company know what its workforce knows? Are they trying to find out?

◇ Are there processes in place that monitor current core competencies and project future requirements? Is your strategic planning based on an explicit understanding of your core competencies and their market value? Is the rate of change of knowledge and the value of competencies in your markets factored into this planning? Are your operations deliberately built around your core competencies, or are they built around established procedures that utilize core competencies inefficiently?

Agility Attributes IV: Knowledge-Driven Enterprise.
See instructions for worksheet 2 on page 253.

Worksheet 5   Agility Attributes IV: *Knowledge-Driven Enterprise*

Date _____ Company Name _____

KNOWLEDGE-DRIVEN ENTERPRISE	IMPORTANCE X OPPORTUNITY = VALUE		
	in your Markets	for your Company	to your Company
Dynamic, Competency-Based Strategic Plan			
Corporate Knowledge Capture Processes			
Expertise-Centered Operations			
Enterprise-Level Performance Metrics			
Open Information Policies			
Open Communication Policies			
Innovative Workforce			
Continuous Education and Training			
Cross-Functional Training			
Internalization of Societal Values			

Reproduced by permission of The Agility Forum
125 Goodman Drive, Bethlehem PA 18015
Tel (610) 758-5510   Fax (610) 694-0542

Scale
1 = Low Importance/Opportunity
3 = Medium Importance/Opportunity
5 = High Importance/Opportunity

ENABLING INFRASTRUCTURE

In the previous chapter we identified interoperability, reconfigurability, and flexibility as generic enabling requirements for supporting the behaviors that an agile enterprise displays. We have already
discussed the roles played by those enablers in a company committed to being a solutions provider and engaging in collaborative
operations. In preparation for the next chapter, we consider the
roles played by these same three enablers in supporting the move
to being an adaptive organization and a knowledge-driven enterprise.

Interoperability

If the same work processes and computer systems were used
throughout an organization, then people and teams from many different parts of that organization would be able to rapidly adapt to
changing requirements, and would be able to manage the knowledge deployed by the people in the organization. Attaining such
uniformity is impossible because of the greatly varied requirements
of different parts of the organization. There is no work process
suited to everyone, and there is no computer system which will
serve all purposes. The practical option is to use different systems,
but to ensure that they are interoperable. This means that in the
whole system, which will necessarily have many different components, each will be able to work with another. They will not be identical, nor will they be integrated into a bigger system which
coordinates between them, but they will be able to work together.

A cross-functionally trained workforce that is routinely reas-
signed to new teams, and to new and transient organizational structures and tasks, needs to be able to produce useful work as quickly
as possible. Creating interoperability at the level of user-technology
interfaces, business practices, and operational procedures, reduces the time necessary for producing useful work.

Facilitator's Warm-Up Questions

◇ Is your company aware of the issues associated with interoperability? Are resources being devoted to implementing interoperability where it would have the greatest impact, for example, in your information system? Are current user-technology interfaces an obstacle to changing work assignments or to smooth cross-departmental teaming? Is the sharing of knowledge hampered by lack of interoperability?

Reconfigurability

In an interprise, the boundaries between departments and teams surround cross-functional clusters of redistributable resources, while points of responsibility change as the customer-centered efforts change.

The concept of teaming is one response to the value in today's competitive environment of customer process-driven operations. Reconfigurability at the organizational level has profound implications for traditional notions of executive power, privilege, and authority. No level of management can be relieved of operational value-creating efforts. Pervasive reconfigurability is incompatible with fiefdoms and fixed concentrations of power, privilege, and authority.

Facilitator's Warm-Up Questions

◇ How routinely can resources from different departments or divisions of your company be configured for a new development effort? Is the identification of individual executives with power over resources an obstacle to implementing an adaptive organization? Is organizational reconfigurability a goal in your company? Have the benefits of organizational reconfigurability been calculated and the risks weighed?

Flexibility

In the interprise, strategic plans continuously evaluate current and required core competencies. At any time, a company needs to support its existing competencies and invest in the acquisition and support of new ones. An interprise should not rely on predictions of the directions in which their products and services will need to evolve, since being part of customers' processes will uncover these needs continuously. Education and training programs, recruitment, decisions about outsourcing, and investment in new technologies are all impacted by unanticipatable requirements derived from the intense interactions with customers, who are themselves constantly confronted with new technologies and opportunities.

Similarly, the interprise has to have flexible policies concerning information sharing and intellectual property rights, outsourcing, partnering, standards to be supported, cooperative relationships with suppliers, and commitments to customers. The ideal customer is one who is a partner in a relationship that is of high mutual value, one in which the customer stimulates the interprise to develop a stream of new, high value and high profit, products and services, both for that customer and for others. This is superbly demonstrated by Ross Operating valves, who say they "mine" the minds of their interactive customers to find new products for their catalogue valve division.

Facilitator's Warm-Up Questions

◇   How flexible are your relationships with customers, suppliers, and partners? How often do you match market developments against your strategic planning, competency requirement assessments, and methods for allocating resources?

◇   Are inflexible compensation and reward mechanisms an obstacle to effective teaming, across departments or division and on multifunction teams?

◇   Do you monitor the evolving mechanisms for outsourcing and partnering? Do you have a mechanism for tracking the evolving value of each of your company's customers, suppliers, and partners?

CHAPTER 25

THE MEASURE
OF SUCCESS

F ollowing identification of your highest priority market forces and the required interprise attributes, this chapter will help you to compile a list of the highest priority action items you need to deal with in order to become an interprise. Real action needs metrics to measure whether efforts are going in the right direction, though the interprise is too young to have generated a standard handbook of operating procedures and metrics. We have, however, identified some of the kinds of metrics that can facilitate your choice of suitable measures for your organization. The objects of the action plan are business processes. A list of these is given below so you can identify which processes are relevant to your business. From this list, you should select no more than five which can be used to run your organization. If you focus on everything, you end up focused on nothing, and it is important to separate the metrics that are necessary from measures which are nice to have but are not vital.

SUGGESTED METRICS FOR AN ACTION PLAN

Select five of the following:

1) New accounting metrics which emphasize flow rates of money through the interprise (chapter 19)

2) Attainable work process velocity and acceleration (chapter 5)

3) The value created by a company for its customers' customers (chapters 13 and 23)

4) The value added on a continuing basis by using products as platforms for information and services, and by making them upgradeable and reconfigurable (chapter 6)

5) The value of individualizability and market fragmentation (chapters 7 and 8)

6) The value of constantly evolving relationships with customers, suppliers, and partners (chapters 4, 11, and 15)

7) The value of the virtual organization model relative to owning the assets needed, or to traditional partnerships (chapter 16)

8) The changing market value of core competencies; that is, the value created by the skills and knowledge of the workforce, as a basis for outsourcing decisions (chapter 9)

9) The value added to the company's bottom-line and/or competitiveness by the skills, knowledge, and initiative of employees (chapters 18 and 20), and appropriate reward mechanisms

10) The proficiency with which the company can change processes, procedures, and organizational structures, reactively and proactively (chapter 24)

11) The speed with which partnerships and supplier relationships can be formed and their operation become effective (chapters 17 and 23)

INTERPRISE MANAGEMENT METRICS

The competitive power of the interprise derives from the network of relationships that it is able to manage to the mutual benefit of the parties interrelated. These relationships—for example, physically dispersed, cross-functional teams within a company, and interactive partnerships with customers and suppliers—create value, but they also create new dependencies, and thus, new kinds of vulnerabilities.

The interprise is a more tightly integrated system than the traditional enterprise. To realize its benefits and protect against its vulnerabilities, managers need access to system-level performance metrics on as close to a real-time basis as possible. For example, operators of the electric utility "interconnects," who control the pooled capacity of numerous utility companies in several states in the U.S., have complete, constantly updated information about the availability and operating cost of every piece of generating equipment in each of the interconnected utilities. This knowledge allows them to match generation to demand at the lowest possible cost to the system as a whole. Individual utilities have their own power plants turned on or off in the best interests of the pool, rather than from their own narrow perspective. The utilities discovered that in the long run, they benefit most individually when they allow such short-run concessions to the group. It is beginning to be appreciated that a similar logic is applicable to the integrated networks of what were until recently sovereign corporate "micro-states." The best interest of these companies also rests on subordinating the pursuit of their exclusive interest to sharing in the greater value-creating capabilities of the integrated, interactive group.

Managers of an interprise need performance measures relevant to a workforce composed of a constantly changing mix of permanent employees, temporary employees, personnel on loan from supplier, partner, and customer companies, and contract workers, including technical professionals and executives. In addition to the metrics listed above, the interprise needs to be able to:

- Assess the relative value and vulnerabilities of utilizing alternative combinations of internal and external resources, ranging from personnel competencies to production equipment.

- Measure the effectiveness with which its mixed workforce is being managed.

INTERPRISE BUSINESS PROCESSES

In chapter 22 you identified the forces driving your business change. In chapters 23 and 24, you selected those enterprise attributes and their enablers that hold the greatest potential for improving your company's competitiveness. Now you can transfer to worksheet 6 the highest impact drivers of change for your company from worksheet 1, and the most valuable change opportunities from worksheets 2 to 5. It is important to consider the impact of these selections on your company's business processes, taking care not to treat these processes in isolation from one another. The following list of business processes should trigger thoughts as to the prioritized list of business processes that you want to deal with in your Action Plan (column 3 of worksheet 6). The worksheets capture, in capsule form, your rationale for change and the primary features of how you believe your company must operate differently if it is to create and sustain competitive advantage.

Worksheet 6

Cumulative Record of Selections

Date _____ Company Name _____

Highest Impact Drivers of Change (from Worksheet 1)	Highest Value Change Opportunities (from Worksheets 2 to 5)	Action Items

BUSINESS PROCESSES FOR ACTION PLAN

Enterprise Management
Strategic Planning
Stakeholder and Partner Relations
Core Competency Analysis
Operations
Physical Resources
Knowledge Assets
Human Resources
Information Systems
Financial Systems
Logistics
Innovation
Risk
Regulatory Compliance

Customer Identification/Creation
Market Research
Joint Product Definition
Product Realization Strategy and Business Plan
Customer Portfolio Management Strategy
Management of Upgrades and Versions
Service and Support

Supplier Management
Sourcing and Partnering Strategies
Product and Process R&D
Operations
Version Control
Information Systems
Logistics

SAMPLE WORKSHEETS COMPLETED BY TWO COMPANIES

The following worksheets were filled out by two companies (their names have been changed) after facilitated group discussions had led to a consensus. The first is a service company, Able County Hospital; the second; a well-known, established, industrial corporation, Universal Manufacturing, Inc. In each case, after reaching agreement on the most threatening market forces and the most valuable opportunities in the interprise model, the group was able to identify the elements of an action plan that they believed could be implemented and would significantly improve their organization's competitiveness.

1.   Able County Hospital

Able County is a medium-size hospital in a stable metropolitan area. Like all health care institutions today, it is under intense pressure from third-party payers and consumers to reduce costs while maintaining a high quality array of constantly expanding, health-related services that go far beyond traditional medical care. At the same time, it is experiencing additional competitive pressure from regional institutions that are attempting to improve their own financial position by drawing away Able's patient population base, attracting Able County physicians to affiliate with them, and by building health service delivery facilities in Able County. Finally, regional and national for-profit health-care providers are exploiting economies of scale and rationing of care to undermine the financial viability of local not-for-profit institutions.

A series of interactive exercises in a two-day workshop with Able County's strategic planning group generated the consensus assessment reflected in worksheets 1 to 5. The highest ranking selections were listed in worksheet 6, on the basis of which the strategic planning group outlined a three-point action plan. Proceeding with this plan would require mapping the changes planned onto the enabling infrastructure requirements (interoperability, reconfigurability, and flexibility) and the business processes affected by the planned changes. At this point, the plan is a pointer to the kinds of change the group believes must be implemented in the near term. Follow-on exercises identified obstacles to be overcome, resources required, metrics needed to consolidate the changes made and to measure their effectiveness, and a time line for accomplishment.

1)   Change current external relationships.

- Identify the distinctive requirements and values of Able County's various customers: patients, third-party payers, affiliated physicians, regional industry, the cit-

izens of Able County, the hospital's board of directors,
and the Able County Board of Commissioners.

- Develop new relationships with these differentiated
  customers, such as regional physicians targeted by rival health-care organizations, based on their distinctive values and their distinctive requirements for
  satisfaction in dealing with Able County.

- Identify opportunities for cooperative relationships
  with rival institutions competing for access to Able
  County's patient base and physician pool.

- Implement a program for cultivating for-profit competitive tactics—the better to stave off competitors
  and to understand the values driving key customers
  and suppliers.

- Explore opportunities for outsourcing functions that
  are clearly not core competencies; for example, laundry services, supplies replenishment, movement of
  personnel and materials among the hospital's physically dispersed facilities, security.

2) **Change current internal relationships.**

- Unite the currently fragmented workforce—divide by
  facility and in some cases by function—into a single
  workforce with a common affiliation to Able County.

- Radically decrease—by a factor of ten—the number
  of job categories, currently numbering in excess of
  300.

- Create an innovative workforce by encouraging and
  rewarding initiative, distributing decision-making,
  and enhancing career development opportunities
  through hospital-funded education and training.

- Create an adaptive work organization by stimulating a cross-functional team approach to work, utilizing the united Able County workforce as a central resource.

- Develop relationships with physicians, nurses, and therapists to enable cross-functional teaming approaches to the organization of work.

- Implement an information system capable of supporting decentralized decision-making, interactive customer and supplier relationships, and an innovative workforce, by providing as comprehensive and open an information environment as is consistent with maintaining security and confidentiality.

3) **Become a provider of integrated healthcare solutions rather than a purveyor of bundles of health care services.**

- As a first step towards creating solutions, Able County Hospital personnel must build relationships with its various customers in order to better understand their current needs and to anticipate new ways of adding value for those customers, for example, by solving problems that customers have come to accept and live with or have not yet identified as problems. At the same time, Able County Hospital must create partnerships with appropriate groupings of customers and suppliers affected by packaging comprehensive, integrated health-care solutions, such as pharmaceutical companies, distributors, physicians, insurers, staff, county health, and financial management personnel.

- Assign customer champions to a prioritized list of Able's key customers. These champions are to be charged with leading team evaluations of customers'

evolving health services needs and concerns, prioritizing these, feeding this information back to hospital
staff, customers, and key suppliers, and developing
win-win solutions and opportunities.

- Develop a process for identifying niche markets for
  health services and for cost-effectively packaging
  Able County competencies to market them.

- Develop performance metrics derived from customer
  feedback.

- Create cross-functional solutions-product realization
  teams, including representatives of key suppliers and
  customers.

Worksheet 1   **Market Forces Assessment**

Date __7/26/99__ Company Name __Able County Hospital__

DRIVERS OF BUSINESS CHANGE	IMPORTANCE in your Markets	X THREAT to your Company	= VALUE to your Company
INTENSIFYING COMPETITION			
Rapidly Changing Markets	5	5	25
Declining Cost of Information	5	5	25
Improving Communication Technologies	3	3	9
Pressure on Costs/Productivity	5	5	25
High Rate of Innovation	3	3	9
Decreasing Time-to-Market	5	3	15
Global Competitive Pressures	1	1	1
FRAGMENTATION OF MASS MARKETS			
Growth of Niche Markets	5	3	15
High Rate of Model Change	1	1	1
Shrinking Product Lifetimes	1	1	1
Shrinking Profit "Windows"	1	1	1
Low Niche Market Entry Costs	5	5	25
COOPERATIVE BUSINESS RELATIONSHIPS			
Inter-Enterprise Cooperation	5	5	25
Interactive Business Relationships	5	5	25
Increasing Outsourcing	3	3	9
Global Sourcing/Marketing/Distribution	1	1	1
Shifting Resource Constraints	5	3	15
Labor-Management Relationships	3	1	3
EVOLVING CUSTOMER EXPECTATIONS			
Individualized Products/Services	5	3	15
Life Cycle Product Support	3	3	9
Rapid Time-to-Market	5	5	25
Rapid Delivery	3	1	3
Changing Quality Expectations	5	5	25
Value of Information/Services	5	3	15

Reproduced by permission of The Agility Forum
125 Goodman Drive, Bethlehem PA 18015
Tel (610) 758-5510 Fax (610) 694-0542

Scale
1 = Low Importance/Threat
3 = Medium Importance/Threat
5 = High Importance/Threat

Worksheet 2   **Agility Attributes I:** *Solutions Provider*

Date __7/26/99__ Company Name __Able County Hospital__

SOLUTIONS PROVIDER ATTRIBUTE	IMPORTANCE X in your Markets	OPPORTUNITY = for your Company	VALUE to your Company
Niche Marketer: High Product Diversity	5	5	25
High New Product Introduction Rate	1	1	1
Frequent Model Changes	1	1	1
Rapid Concept-to-Cash	5	3	15
Cost-Effective Low-Volume Producer	5	1	5
Production to Order	1	1	1
Individualizable Products/Services	5	1	5
Life Cycle Design Methodology	1	1	1
Open Architecture Product Design Philosophy	1	1	1
Life Cycle Product/Customer Support	5	5	25
Information/Services-Rich Products	5	5	25
Enduring, Proactive Customer Relationships	5	3	15
Proactive Marketplace Change Agent	5	5	25
Solution-Based Marketing Policies	5	5	25
Extraordinary Quality Standards	5	5	25
Market-Opportunity Pulled Production	5	3	15

Scale
1 = Low Importance/Opportunity
3 = Medium Importance/Opportunity
5 = High Importance/Opportunity

Worksheet 3   Agility Attributes II: *Collaborative Operations*

Date _7/26/99_ Company Name _Able County Hospital_

COLLABORATIVE OPERATIONS	IMPORTANCE in your Markets	X OPPORTUNITY for your Company	= VALUE to your Company
Cooperation = Operations Strategy of First Choice	5	3	15
Concurrent Operations	1	1	1
Integrated Product and Process Development	3	3	9
Integrated Comprehensive Enterprise Processes	5	3	15
Interactive Customer/Supplier Relationships	5	5	25
Virtual Organization Partnering	3	5	15
Electronic Commerce Operability	5	5	25
Proactive Information Sharing Policies	3	3	9

Scale

Reproduced by permission of The Agility Forum
125 Goodman Drive, Bethlehem PA 18015
Tel (610) 758-5510   Fax (610) 694-0542

1 = Low Importance/Opportunity
3 = Medium Importance/Opportunity
5 = High Importance/Opportunity

Worksheet 4 Agility Attributes III: *Adaptive Organization*

Date _7/26/99_ Company Name _Able County Hospital_

ADAPTIVE ORGANIZATION	IMPORTANCE in your Markets	X OPPORTUNITY for your Company	= VALUE to your Company
Motivational Management Philosophy	5	5	25
Coordinated, Decentralized Decision-Making	5	3	15
Adaptive and Reconfigurable Physical Resources/Processes	3	3	9
Adaptive Work Organization	5	5	25
Timely, Opportunity-Driven Organization	3	3	9
Change Proficient Organization	3	5	15
Adaptive Information System	5	5	25
Distributed Business/Production Processes	3	1	3

Reproduced by permission of The Agility Forum
125 Goodman Drive, Bethlehem PA 18015
Tel (610) 758-5510   Fax (610) 694-0542

Scale
1 = Low Importance/Opportunity
3 = Medium Importance/Opportunity
5 = High Importance/Opportunity

Worksheet 5   Agility Attributes IV: *Knowledge-Driven Enterprise*

Date _7/26/99_ Company Name _Able County Hospital_

KNOWLEDGE-DRIVEN ENTERPRISE	IMPORTANCE In your Markets	X OPPORTUNITY for your Company	= VALUE to your Company
Dynamic, Competency-Based Strategic Plan	5	3	15
Corporate Knowledge Capture Processes	5	3	15
Expertise-Centered Operations	5	1	5
Enterprise-Level Performance Metrics	5	5	25
Open Information Policies	3	3	9
Open Communication Policies	5	3	15
Innovative Workforce	5	5	25
Continuous Education and Training	5	5	25
Cross-Functional Training	3	3	9
Internalization of Societal Values	5	1	5

Reproduced by permission of The Agility Forum
125 Goodman Drive, Bethlehem PA 18015
Tel (610) 758-5510   Fax (610) 694-0542

Scale
1 = Low Importance/Opportunity
3 = Medium Importance/Opportunity
5 = High Importance/Opportunity

Worksheet 6

Cumulative Record of Selections

Date __7/26/99__ Company Name __Able County Hospital__

Highest Impact Drivers of Change (from Worksheet 1)	Highest Value Change Opportunities (from Worksheets 2 to 5)	Action Items
Rapidly Changing Markets	Niche Marketer	(1) Change external relationships.
Declining Cost of Information	Life Cycle Customer Support	(2) Change internal relationships.
Pressure on Costs	Information/Service-Rich Products	(3) Become a solutions-provider.
Low Niche Market Entry Costs	Proactive Marketplace Change Agent	(See page 281 for details.)
Inter-Enterprise Cooperation	Solutions-Based Marketing	
Interactive Business Relationships	Extraordinary Quality	
Rapid Time to Market	Interactive Relationships	
Evolving Quality Expectations	Electronic Commerce	
	Motivational Management Philosophy	
	Adaptive Work Organization	
	Adaptive Information System	
	Enterprise-Level Metrics	
	Innovative Work Force	
	Continuous Education and Training	

2. Universal Manufacturing, Inc.

Universal Manufacturing is a peripheral division of a global corporation. Its revenues are generated overwhelmingly from the sale of a narrow range of standardized products to a global customer base concentrated in one segment of a rapidly expanding consumer service industry. Competition is fierce, customer loyalty is nonexistent, and Universal's sales force is "borrowed" from another division of the corporation where compensation is keyed to unit equipment sales rather than to building value-adding relationships over time.

A series of two-day interactive exercises with Universal's senior management team generated the ranking of drivers and opportunities contained in worksheets 1 to 5. These assessments reflect opportunities in Universal's markets for growth by selling solutions in spite of intense pressure to reduce costs and compete on unit price. The highest value entries were recorded in worksheet 6. Extensive facilitated discussion of how to exploit the opportunities listed in worksheet 6 resulted in a strong group commitment to a four-point action plan that was checked for justification against the highest impact market drivers.

1) **Implement interactive relationships—aggressively, proactively, and innovatively managed—with customers and suppliers.**

- For a key supplier whose performance is a major factor in quality problems and long lead times, offer space within the company's own facilities for manufacturing the components supplied, using the supplier's personnel and equipment, with all infrastructure requirements and a share of the moving expenses covered by Universal Manufacturing. Offer a product redesign to help the supplier meet desired quality and lead time goals. If the supplier refuses and does not offer an alternative with comparable quality improvement and lead time reduction outcomes, contract with another supplier in spite of the near-term pain involved.

- Create pre-sale and post-sale support systems for customers as well as for Universal's own sales force, which in the past has been rewarded for single instance sales, rather than for creating value-adding relationships over time.

- Place technically knowledgeable people into sales channels as resources to customers and to Universal's sales force.

- Invite selected customers and suppliers to participate in new product development.

- Implement knowledge-capture processes centered first on Universal's current customers, and then expand to suppliers and markets (potential customers).

- Create cross-functional accountability teams for supplier performance and customer satisfaction.

- Determine the responsibility of Universal for its problems with suppliers and identify Universal-led initiatives for overcoming these problems.

2)   **Develop information and service products.**

- Package predefined levels of service as products.

- Sell software packages and services, which are currently bundled with machinery at little or no profit, as products in their own right.

- Develop a solutions-competent sales force by training the existing product-oriented sales force as well as recruiting new personnel.

- Develop software tools for customers, for Universal's own sales force, and for suppliers that will enhance the value of doing business with Universal by making it easier, faster, and less expensive.

3) Commit to extraordinary quality standards.

- Differentiate the definition of "quality" and its measurement by context. That is, distinguish what is meant by quality, the market value of that definition, and appropriate quality metrics by context. For example, consider quality as a property of the processes used to produce products; quality as a property of Universal's products and services as delivered; and quality as a property of the customer's experience with using Universal's products and services.

- Incorporate identification of what it will take to exceed an individual customer's expectations into the ordering process.

- Create processes at Universal that force personnel to think through the value of products and services from the customer's perspective.

- Open service records to customers as a means of showing the reliability of Universal's products. This will require creating and maintaining easy access to comprehensive and current service records.

- Build relationships with the users of Universal's products at customers' sites in order to understand better how to add value for them.

4)   **Create an innovative workforce supported by a motivational (as opposed to a command-and-control) management philosophy and adaptive work organization
structures.**

NOTE WELL that this clustering of different agile enterprise attributes into a single, compound action item illustrates how the translation of opportunities into an action
plan leads to assessing the functional linkages among the
attributes.

• Compile a total workforce competency inventory by
identifying skills and skills training.

• Use this inventory to create skills-based project
teams at Universal, treating the total workforce as a
single resource pool to be utilized as required by market opportunities.

• Distribute decision-making more widely and deeply,
but implement mechanisms for coordinating decision-making so as to enhance enterprise performance
without recentralization of authority.

• Rotate management job assignments every three to
five years.

• Challenge all existing processes and procedures, but
always in the context of the collateral processes and
procedures to which they are functionally coupled.

SUMMARY

These sample assessments are intended to provoke your thinking
about how the interprise approach to competition could benefit
your company. As we said in our earlier book, there is no formula
for transforming a company into an aggressive, market-opportunity
exploiting interprise, but there is also no alternative, except competitive marginality.

Worksheet 1   **Market Forces Assessment**

Date _2/17/99_ Company Name _Universal Mfg._

DRIVERS OF BUSINESS CHANGE	IMPORTANCE in your Markets	X THREAT to your Company	= VALUE to your Company
INTENSIFYING COMPETITION			
Rapidly Changing Markets	5	5	25
Declining Cost of Information	5	1	5
Improving Communication Technologies	3	3	9
Pressure on Costs/Productivity	5	5	25
High Rate of Innovation	5	3	15
Decreasing Time-to-Market	3	1	3
Global Competitive Pressures	5	5	25
FRAGMENTATION OF MASS MARKETS			
Growth of Niche Markets	5	3	15
High Rate of Model Change	3	3	9
Shrinking Product Lifetimes	3	3	9
Shrinking Profit "Windows"	5	5	25
Low Niche Market Entry Costs	5	3	15
COOPERATIVE BUSINESS RELATIONSHIPS			
Inter-Enterprise Cooperation	3	3	9
Interactive Business Relationships	3	3	9
Increasing Outsourcing	3	1	3
Global Sourcing/Marketing/Distribution	5	5	25
Shifting Resource Constraints	3	1	3
Labor-Management Relationships	3	3	9
EVOLVING CUSTOMER EXPECTATIONS			
Individualized Products/Services	3	3	9
Life Cycle Product Support	5	3	15
Rapid Time-to-Market	5	5	25
Rapid Delivery	5	5	25
Changing Quality Expectations	5	5	25
Value of Information/Services	5	5	25

Reproduced by permission of The Agility Forum
125 Goodman Drive, Bethlehem PA 18015
Tel (610) 758-5510 Fax (610) 694-0542

Scale
1 = Low Importance/Threat
3 = Medium Importance/Threat
5 = High Importance/Threat

Worksheet 2   **Agility Attributes I:** *Solutions Provider*

Date _2/17/99_   Company Name _Universal Mfg._

SOLUTIONS PROVIDER ATTRIBUTE	IMPORTANCE X OPPORTUNITY = VALUE		
	in your Markets	for your Company	to your Company
Niche Marketer: High Product Diversity	5	3	15
High New Product Introduction Rate	5	3	15
Frequent Model Changes	1	1	1
Rapid Concept-to-Cash	3	3	9
Cost-Effective Low-Volume Producer	5	1	5
Production to Order	5	1	5
Individualizable Products/Services	3	3	9
Life Cycle Design Methodology	3	1	3
Open Architecture Product Design Philosophy	3	1	3
Life Cycle Product/Customer Support	5	3	15
Information/Services-Rich Products	5	5	25
Enduring, Proactive Customer Relationships	5	5	25
Proactive Marketplace Change Agent	5	5	25
Solution-Based Marketing Policies	5	5	25
Extraordinary Quality Standards	5	5	25
Market-Opportunity Pulled Production	3	1	3

Scale
1 = Low Importance/Opportunity
3 = Medium Importance/Opportunity
5 = High Importance/Opportunity

THE MEASURE OF SUCCESS

Worksheet 3   Agility Attributes II: *Collaborative Operations*

Date __2/17/99__   Company Name __Universal Mfg.__

COLLABORATIVE OPERATIONS	IMPORTANCE in your Markets	X OPPORTUNITY for your Company	= VALUE to your Company
Cooperation = Operations Strategy of First Choice	3	5	15
Concurrent Operations	3	3	9
Integrated Product and Process Development	3	5	15
Integrated Comprehensive Enterprise Processes	5	3	15
Interactive Customer/Supplier Relationships	5	5	25
Virtual Organization Partnering	3	5	15
Electronic Commerce Operability	5	3	15
Proactive Information Sharing Policies	3	5	15

Reproduced by permission of The Agility Forum
125 Goodman Drive, Bethlehem PA 18015
Tel (610) 758-5510   Fax (610) 694-0542

Scale
1 = Low Importance/Opportunity
3 = Medium Importance/Opportunity
5 = High Importance/Opportunity

Worksheet 4   **Agility Attributes III:** *Adaptive Organization*

Date __2/17/99__   Company Name __Universal Mfg.__

ADAPTIVE ORGANIZATION	IMPORTANCE in your Markets	X OPPORTUNITY for your Company	= VALUE to your Company
Motivational Management Philosophy	5	5	25
Coordinated, Decentral.zed Decision-Making	5	5	25
Adaptive and Reconfigurable Physical Resources/Processes	3	5	15
Adaptive Work Organization	3	5	15
Timely, Opportunity-Driven Organization	5	5	25
Change Proficient Organization	3	5	15
Adaptive Information System	5	3	15
Distributed Business/Production Processes	3	3	9

Scale
1 = Low Importance/Opportunity
3 = Medium Importance/Opportunity
5 = High Importance/Opportunity

Worksheet 5   Agility Attributes IV: *Knowledge-Driven Enterprise*

Date **2/17/99** Company Name *Universal Mfg.*

KNOWLEDGE-DRIVEN ENTERPRISE	IMPORTANCE in your Markets	X OPPORTUNITY for your Company	= VALUE to your Company
Dynamic, Competency-Based Strategic Plan	3	5	15
Corporate Knowledge Capture Processes	5	5	25
Expertise-Centered Operations	3	5	15
Enterprise-Level Performance Metrics	5	3	15
Open Information Policies	3	5	15
Open Communication Policies	3	5	15
Innovative Workforce	5	5	25
Continuous Education and Training	3	5	15
Cross-Functional Training	3	5	15
Internalization of Societal Values	1	3	3

Reproduced by permission of The Agility Forum
125 Goodman Drive, Bethlehem PA 18015
Tel (610) 758-5510   Fax (610) 694-0542

Scale
1 = Low Importance/Opportunity
3 = Medium Importance/Opportunity
5 = High Importance/Opportunity

Worksheet 6

Cumulative Record of Selections

Date __2/17/99__ Company Name __Universal Mfg.__

Highest Impact Drivers of Change (from Worksheet 1)	Highest Value Change Opportunities (from Worksheets 2 to 5)	Action Items
Rapidly Changing Markets	Information/Service-Rich Products	(1) Implement Interactive Customer/Supplier Relationships
Pressure on Costs/Productivity	Proactive Customer Relationships	(2) Develop Information and Service Products
Global Competitive Pressure	Proactive Change Agent	(3) Commit to extraordinary quality standards
Shrinking Profit Windows	Solutions-Based Marketing	(4) Create an innovative workforce
Global Sourcing/Marketing	Extraordinary Quality	(See page 291 for details.)
Rapid Time-to-Market	Interactive Customer/Supplier Relationships	
Rapid Delivery	Motivational Management Philosophy	
Changing Quality Expectations	Coordinated Decentralized Decision Making	
Value of Information/Services	Opportunity Driven Organization	
	Knowledge Capture Processes	
	Innovative Workforce	

Reproduced by permission of The Agility Forum, 125 Goodman Drive, Bethlehem PA 18015 • Tel (610) 758-5510 Fax (610) 694-0542

REFERENCES AND SUGGESTED READINGS

Amelio, Gil and Simon, William, *Profit from Experience*, Van Nostrand Reinhold, 1996.

Association for Manufacturing Excellence, Wheeling, Illinois, *Target* magazine, especially Volume 11, Number 4, July-August 1995.

Bennis, Warren and Townsend, Robert, *Reinventing Leadership*, Simon and Schuster, 1991.

Champy, James, *Reengineering Management*, Harper Collins, 1995.

Chandler, Alfred D., Jr., *The Visible Hand*, Cambridge: Harvard University Press, 1977.

Collins, James and Porras, Jerry, *Built to Last*, Harper Business, 1991.

Cooper, Robin, and Chew, W. Bruce, *Control Tomorrow's Costs Through Today's Designs*. Harvard Business Review, Jan.–Feb. 1996, pp. 88-97.

Cronin, Mary J., *Doing More Business on the Internet*, Van Nostrand Reinhold, 1995.

Drucker, Peter Ferdinand, *Post Capitalist Society*, Harper Business Press, 1993.

Gascoyne, Richard J., *Corporate Internet Planning Guide: Aligning Internet Strategy with Business Goals*, Van Nostrand Reinhold, 1996.

Goldman, Steven L., Nagel, Roger N. and Preiss, Kenneth, *Agile Competitors and Virtual Organizations: Strategies for Enriching The Customer*, Van Nostrand Reinhold, 1995.

Goldman, Steven L. and Preiss, Kenneth, Editors; Nagel, Roger. N. and Dove, Rick, *Principal Investigators, with 15 industry executives, 21st Century Manufacturing Enterprise Strategy: An Industry-Led View*, 2 volumes, Iacocca Institute at Lehigh University, Bethlehem PA, 1991.

Goldratt, Eliyahu M., *The Goal: a Process of Ongoing Improvement*, North River Press, 1984.

Goldratt, Eliyahu M., *The Haystack Syndrome*, North River Press, 1990.

Goranson, H. Ted, *A Whale of a Tale—A Historical Perspective of Virtual Enterprise*, PA 96–2, Agility Forum, Bethlehem, PA, 1996.

Hall, Robert, *The Soul of the Enterprise* New York: Harper Business, 1993.

Hamel, Gary and Prahalad, C.K., *Competing for the Future*, Harvard Business School Press, 1994.

Hayes, Robert H., Wheelwright, Steven C., Clark, Kim B., *Dynamic Manufacturing—Creating The Learning Organization*, The Free Press, 1988.

Imai, Masaaki, *Kaizen—The Key to Japan's Competitive Success*, The Kaizen Institute, Ltd. 1986.

Johnson, Thomas H. and Kaplan, Robert L., *Relevance Lost*, Cambridge, MA: Harvard Business Press, 1987.

Johnson, Thomas H., *Relevance Regained*, New York, Free Press, 1992.

Lewis, Jordan D., *The Connected Corporation*, The Free Press, 1995.

Moore, James F., *The Death of Competition*, HarperBusiness, 1996.

Nonaka, Ikujiro and Takeuchi, Hirotaka, *The Knowledge—Creating Company*, Oxford University Press, 1995.

Patterson Marvin L., *Accelerating Innovation*, Van Nostrand Reinhold, 1993.

Preiss, Kenneth, Leary, Jennifer L. and Jahn, William J. II, editors, *Handbook for Virtual Organization: Tools for Management of Quality, Intellectual Property and Risk & Revenue Sharing*, Knowledge Solutions Inc., Bethlehem Pa, 1996.

Quinn, James Brian, *Intelligent Enterprise*, New York, Free Press, 1992.

Savage, Charles, *Fifth Generation Management*, Butterworth-Heinemann, 1996.

Senge, Peter, *The Fifth Discipline: The Art and Practice of the Learning Organization*, Doubleday, 1996.

Shulman, Ron, *Rhone-Poulenc Uses Profit Velocity to Pave the Way to Agility*, Agile Enterprise Journal, John Wiley, Fall 1996.

Slywotzky, Adrian J., *Value Migration*, Harvard Business School Press, 1996.

Thurow, Lester, *The Future of Capitalism: How Today's Economic Forces Shape Tomorrow's World*, William Morrow, 1996.

Turney, Peter B. B., *Common Cents: The ABC of Performance Breakthrough*, Cost Technology, Hillsboro, OR, 1991.

Treacy, Michael and Wiersma, Fred, *Discipline of Market Leaders*, Addison-Wesley, 1995.

Warnecke, Hans-Jurgen, *The Fractal Company: A Revolution in Corporate Culture*, Springer Verlag, 1993.

Warnecke, Hans-Jurgen, Huser, Manfred and Kaum, Ralf, *A Matter of Scale(s)—Made-to-Order Weighing Machines Bring Success to Mettler-Toledo*, Agile Enterprise Journal, John Wiley, Fall 1996.

INDEX

A

Able County Hospital sample worksheets, 281-290
Activity-Based Costing (ABC), 198-200
Adams, Bill, 104
added-value
    See customer; enrichment dimension; supplier
Agile Web of Pennsylvania
    business networking, 162-163
    knowing the customer, 104
    values and ethics, 171-173
    virtual organization, 192
agility attribute enablers, 260-261
Agility Attributes I: Solutions Provider worksheet, 254
Agility Attributes II: Collaborative Operations worksheet, 257
Agility Attributes III: Adaptive Organization worksheet, 267
Agility Attributes IV: Knowledge-driven Enterprise worksheet, 271
Agility Forum, Iacocca Institute of Lehigh University
    as contributor to interprise model, 238
    interprise model context, 243
    team skills, 184-185
air traffic systems, 35-36
airline industry, 159
Allied Signal, 7
Anderson Consulting, 94
Andrew, James P., 94
Apple Computer, 56, 105
Asea Brown Boveri Canada (ABB)
    partnering, 22
    self-motivated teams, 215
AT&T
    Lucent Technology, 85
    product fusion, 57

B

Bailyn, Lotte, 219
beer market, 78-79
Bergen Advertising, 56
blitzkrieg military strategy, 15
BMW plant, Regensberg, 188

Bossidy, Larry, 7
British Telecom, 159
business process
    See also customer business process; interprise
    action plans, 278-294
    changes, 3-4
    chart, 5
    between companies, 33-34
    customer-pulled process, 36, 132
    fragmenting markets, 82-83
    integrating with customers and suppliers, 103-105, 133
    as interactive processes, 16
    linkage dimension, 117-119
    personalized products, 44
    time compression and, 45-51
    using information technology, 57-59

C

caller ID, 124
Cambridge Technology Partners (CTP), 133-134
Cameron, Russ, 45-46
Carlsson, Jan, 187, 225-226
Caterpillar, 97
Cato the Elder, 18
Chrysler
    combining World Wide Web and multimedia technology, 94
    development time, 44
    SJ6 engine, 152
    supplier input on design, 140
Churchill, Winston, 231
Cincinnati Milacron, 137
CNN, self-motivated teams, 215
Coca Cola, 59
collaboration, 255-257
company-employee relationships, 221-223, 265
competitiveness
    advantages of relationships, 102-104, 111-114
    assessing opportunities, 252-253
    changing environment and, 6
    intensifying, 243

305

people as assets, 216-217
  response to, 15
  through customization, 68-71
  time compression advantages, 41-42
  worldwide, 13-14
constant money flow, 204
consumer
  defined, 67
  feedback from, 90-93
  measurable enrichment, 131-132
cooperation
  See also partnering
  advantages, 7, 11-12
  feasibility, 258
  modalities of, 167
  production relationships, 246-247
core competencies, 171-173
corporate pyramid, 4, 5
cost accounting
  Activity-Based Costing (ABC), 198-200
  allocation methods, 195-198
  80/20 rule, 200
  money flow method, 201-209
  paint method, 195-195
  target costing, 200-201
  time-based costing, 210-212
  traditional methods, 191-194
Crutchfield, Edward, 95
Cumulative Record of Selections worsheet, 279
Custom Foot, shoe company, 24
"Custom Publishing," 70
customer
  adapting to, 265-266
  defined, 67
  expectations, 247-248
  improving bottom line, 126-129
  personal context concept, 102-104
  personalizing service for, 66-67
  point of view, 101-104
  value, 26-30, 112-114, 136-137
customer business process, 17-18, 21
customer enrichment, 123-125, 132, 144
customer information systems, 71
customer satisfaction
  See also competitiveness; consumer
  connecting with customers, 12-13
  interactive relationships, 45-47, 117-119, 150
  mass customization, 24-25
  strategic relationships, 21-23
customer-focused organization, 103
"customerizing" process, 16
customer-pulled process, 36, 132
customization, 68-71

customized solutions, 67-70, 111, 129
  See also mass customization
D
Deutsche Telecom
  and service industry, 87-88
  virtual organization, 159
Dickman, Harry, 230
Direct Inward Dialing, 124
direct satellite systems (DSS), 146
downsizing
  low morale and, 107-108
  military forces, 96-97
  old business environment, 7
Duignan, Henry, 45-46
DuPont
  virtual organization, 146
  work-life programs, 107-108
E
e-Coupons, University of Ann Arbor, Michigan, 92
EDS (electronic information and data systems company), 135-136
Edwards, Jim, 220
80/20 rule, 200
electronic communication
  financial services, 89-90
  global marketing, 85-86
  information and consumer feedback, 90-93
  product servicing and maintenance, 87-88
electronic data interchange (EDI), 153
Electronic Share Information (ESI) Ltd., 90
electronics industry survey, reward-sharing methods, 144
employee morale, 107-108
enrichment dimension
  measurable, 126-129, 131-132
  non-measurable, 122-125
  supplier to customer, 115-116
enterprise-level attributes
  adapting to change, 263-274
  market opportunities, 252-261
entrepreneurship, 218
Estes, Pete, 233
ethics
  See trust
  and Agile Web, 171
  statement of, 172-173
F
Fahrzeugausstattung GmbH, PFA Planungsund Produktionsgessellschaft Fur, Germany, 180
Federated Flowers
  and mindshare, 92-93
  partnering, 104
FedEx Web site, 111

Fine, Charles, 243
First Direct Bank
  communication-based service, 89
  customer information systems, 71
First Union, 95
fixed prices, 117
flexibility, 260-261, 274
"fly by wire" cars, 61
Ford Motor Company, 91
Fortune magazine, 53, 56
France Telecom, 159
Franklin, Nick, 194-195
frequent flyer miles, 123
fusion products, 61-62
fuzzy value, 123-125
G
gambling, 102
Gateway, 81-82
General Electric
  simplifying processes, 80
  Tiger Teams, 180
General Motors
  cost accounting methods, 191, 194-195
  Delphi Saginaw Steering Systems, 36-37,
    169-170
  leadership, 233
Generally Accepted Accounting Principles
  (GAAP), 195
German railways, 180
global alliances, 159
goodwill, 123-125
Goodyear
  customer relationships, 140
  fragmented tire market, 78, 83
  logistics system, 56
  tire servicing, 139-140
groupware information products, 153
H
halving-time, 48-51
Hamilton Bancshares, 89
hardware
  See product platform
Harley-Davidson
  customer perceptions of product, 152
  customer-supplier relationships, 141
  multimedia marketing, 93-94
  product marketability, 251-252
Hewlett-Packard Corporation
  employee empowerment, 230
  organizational efficiency, 23
  product lifetime, 245
Holden Evans, naval contractor, 194
Hughes Electronics, 146
I
Iacocca, Lee, 152
IBM Corporation
  building linkages, 151

multidiscipline task forces, 158
information technology (IT), 57-59
  See also technology
Ingersoll Rand, 49
InstyPrint, 18
intellectual property rights, 14, 159,
  166-167, 170
intelligent airbag development, TRW, 61-62
interoperable systems, 258-259, 272-273
interprise
  See also business process; leadership
  action plan, 240, 242-249
  adapting to change, 263-274
  beneficial/profitable relationships, 111-112
  characteristics, 239-240
  chart, 6
  company-employee relationships, 221-223
  core competencies, 171-174
  creating, 237-238
  defined, 4
  enabling customer satisfaction, 26-31
  entrepreneurship, 218-220
  fusing product+service+information,
    55-58, 61
  generic model, 238-239, 241-242
  improving customer's business, 131-134
  interactivity, 179-181, 234-235
  linking businesses, 35-39, 112-114, 117-119
  management metrics, 277-278
  market opportunities, 252-261
  relationships, 139-141
  strategic partners, 104-108
  success factors, 8, 16, 45
  using time compression, 45-51
Interprise Relationship Model, 112-119
InterRegio passenger cars, German
  railways, 180
Iscar Company
  employee empowerment, 229-230
  global business transactions, 86-87
J
J. Walter Thompson/OnLine, 91-92
Japanese Karetsu, 143
JM Mold, company, 118
John Deere, seeders, 252
Johnson and Johnson, Vascular Access
  Division, 49
Johnson Controls, 17, 132
joint ventures, 145
just-in-time suppliers, 141
  See also suppliers
Juval Bar-On, 230
K
Kefalas, Paul, 22
Key Systems, 94
Keyes, James, 17
Kinko's, 18

knowledge-driven enterprise, 266-271
L
Landis and Gear, 115, 136
leadership
　See also management
　challenges, 231-234
　characteristics, 233-234
　compared to management, 228
　employee responsibilities, 229
　empowerment and, 229-231
　interprise responsibilities, 226-227
　multidimensional complexity, 234-235
　philosophies, 225-226
legacy problems, 258-259
Lengyel, Alec, 180
Levi Strauss, 67, 75
Lexus, 16
lifetime employability, 220-221
linkage dimension
　described, 117-119
　goals, 149
　measurable relationship effectiveness,
　　151-153
　naturally occurring linkages, 150
　virtual teams, 153
Lockheed, 146
logistics, international trade, 96-97
Lord Abbett & Co., 8
Lotus Notes, 153
Lucent Technology, 85, 104
Lufthansa, 87-88
M
Machiavelli, leadership philosophy, 225
Mack Trucks, 54, 72
magazine publishers, 136-137
Maginot Line, 15
management
　See also leadership
　adaptive approach, 264
　challenges, 6
　evaluating trust gaps, 174-176
　knowledge inventory, 269-271
　operational techniques, 14-15
　organizational complexity and, 230-231
　point of view exercise, 106
　work-life programs, 107-108
manufacturers
　See suppliers
manufacturing resource planning system
　(MRP), 127
Market Forces Assessment worksheet, 249
market fragmentation
　expanding markets, 79-80
　fragmenting product lines, 75-79
　mass markets, 244-245
　problems with, 79
　simplifying first, 80-82

mass customization
　customer satisfaction, 24-25
　personalizing products, 67-70
　time compression advantages, 41-42
　using technology, 70-72
mass production
　changes, 23-24, 34
　high cost of, 66
　old model, 101-102
Matthew Outdoor Advertising Inc., 94-95
Mayakawa
　customer enrichment, 136
　using product platform, 59-60
MCI, 159
Mettler, cross-functional teams, 183
Mills, Steve, 182-184
mindshare, 82, 90-93
money flow cost accounting method,
　201-209
Motorola
　job security, 220
　legal counsel, 234
　Schedule Sharing System, 37-38, 152
Mt. Sinai Medical School, 231
multimedia marketing, 93-95
mutually beneficial relationships, 111, 112
N
Nantucket whalers, 158-159
National Insurance Crime Bureau, 96
new product technologies, acceleration
　rate of, 37
Nike
　increasing profits, 83
　market fragmentation, 82
　providing solutions, 252
　sneakerization, 75-76
Nissan, 125
Nordberg, E. Wayne, 8
not-for-profit customers, 126
NuCor Steel, 184
O
Olympus Optical Company, 201
order fulfillment systems, 57
Oregon State prison, 96
organization chart, 4
organizational complexity, 230-231
organizational infrastructure, 256, 272-274
Otis Elevator, 58-59
OTISLINE, 58-59
P
Packard, Dave, 230
partnering
　beneficial/profitable relationships, 111-112
　business enhancement, 134-137
　importance of, 104-105
Personal Journal, 70
"Personal Pair" jeans, 67

personalized goods/services
  See mass customization
Pierron, Joe, 96
"Pittrack" card, 102
price-follows-value arrangements, 143, 144
Prince plant, Michigan, 220-221
"Prison Blues" clothing, 96
process-focused management, 4
product life cycle, 76-79, 245
product platform
  advantages of, 55-56
  defined, 53
  developing, 62-63
  expanding business, 59-60
  using information technology, 57-59
profit-driven customers, 126
"programs of the month," 15
"push" principle, 131-132
Pyfrom, Rick, 49
Q
quick response companies, 50
quick response manufacturing, 44
Quick Response system, 41, 68
R
rapid processing
  See time compression
reconfigurability
  cross-functional communication, 273
  information systems, 259-260
reengineering
  to provide customer value, 136-137
  shortcomings, 6-7
Remmele Engineering Inc., 217
response time
  See time compression
reward dimension
  customer reward/payment, 116-117
  interactivity advantages, 142
  joint ventures, 145
  in special relationships, 140
Rhône-Poulenc Corporation, 211
Roos, Dan, 216
Ross Operating Valves
  customizing, 69
  expanding profitability, 111
  mutually beneficial relationships, 75
  product marketability, 251
  using time compression, 45-47
RossFlex systems, 46
R.R. Donnelly Company
  building linkages, 154
  customer relationships, 8, 140
Runkle, Don, 169, 191
S
SAAB cars, 152
SAS (Scandinavian Airline System)
  goals, 187

leadership, 225-226
Saturn cars, 152
Schrock Cabinet Company, 182-184
Scitex systems, 57-58
self-managing teams
  advantages, 187-188
  coaching and empowering, 182-184
  compared to committees, 181-182
  corporate responsibilities, 219-223
  defined, 179-181
  guidelines, 185-187
  individual contributions, 215-216
  interacting work processes, 63
  nurturing people, 217-218
  personality considerations, 184-185
7 Eleven, 150
shared risk, 116-117
shipbuilders, 48
Sims, James K., 131-134
"sneakerization," 75-79
  See also market fragmentation
Solectron
  Baldrige quality award, 182
  P&L statements, 184
Springfield Remanufacturing, 184
Sprint, 159
stakeholder relationships, 105-108
Steelcase Furniture Company, 26-28
success
  interprise management metrics, 277-278
  metrics for action plan, 276
  priority action items, 275
suppliers
  See also trust
  adding customer-perceived value, 123-125
  communication networks, 96-97
  controlling retail distribution, 33, 37-38
  customer needs and profits, 112
  customer-supplier relationship, 102-105
  enrichment dimension, 115-116
  improving customer's business, 126-129
  rewarding, 139
  sharing information, 141
Swatch watches, 77, 81, 83
T
target costing, 200-201
teams
  See self-managing teams
technology
  See also information technology (IT)
  in mass customization, 70-72
  new product technologies, 37
Texas Instruments
  Defense and Electronic Systems Division, 181-182
  job security, 220
Textile and Clothing Technology

Corporation
  mass customizing, 68-69
  time compression, 41
The New York Times, 70
The Wall Street Journal, 70, 75
3-M magnetic tape plant, 232
Tiger Teams, 180
time compression
  advantages of, 41-45
  in an interprise, 49-51
  changing business process, 45-48
time-based costing, 210-212
Timme, Otto, 187-188
Toshiba computers, 252
Toyota
  development time, 44
  rewarding managers, 183
travel agencies, 159
Trilogy, software company, 94
trust
  See also ethics
  enabling customer relationships, 144
  ethics statement, 172-173
  gaps, 174-176
  in relationship-based business
  environment, 169-171
  supplier agreements, 170
TRW, intelligent airbag development, 61-62
U
U.K. Fine Chemicals, 162
Unisys
  business linking, 151
  customer business processes, 16
  customer enrichment, 132
Universal Manufacturing, Inc. sample
  worksheets, 291-294
University of Ann Arbor, Michigan, eCoupons,
  92
U.S. Robotics, 245
V
value
  See customer; enrichment dimension;
  supplier
value-adding process, 43-44
value-based pricing, 117
Van Nostrand Reinhold, 124
variable money flow, 204
Viola, Donn, 54
Virgin Airlines, 125
virtual organizations
  advantages, 165-166
  characteristics, 158-161
  forming webs, 162-164
  legal form of, 166-167
  opportunities, 145-146
  relationships, 105, 157-158
virtual teams, 153

Volkswagen, 187-188
W
Wall Street, 38
Wal-Mart
  electronic linkage with suppliers, 152
  enrichment process, 116
  win-win-win business philosophy, 33-34
Weaver, Terry, 17
web
  See virtual organizations
Welch, Jack, 80
Wertheimer, Stef, 229-230
Whitney, Eli, 13
win-win relationships, 149
work-life program, 107-108
worksheets
  Agility Attributes I: Solutions Provider, 254
  Agility Attributes II: Collaborative
  Operations, 257
  Agility Attributes III: Adaptive
  Organization, 267
  Agility Attributes IV: Knowledge-driven
  Enterprise, 271
  for business process action plans, 279,
  285-290
  Market Forces Assessment worksheet, 249
World Wide Web, 93, 94
worldwide competitiveness, 13-14
  See also competitiveness; electronic
  communication
X
Xerox Corporation
  good work environments, 219
  integrating with customers, 11-12
Y
Young, John, 23

ABOUT THE AUTHORS

K ENNETH PREISS is an authority on international industrial competitiveness. Over a 35-year career, he has led projects from enterprise strategy and transformation to technical development. He has held leadership roles in defense and industrial projects in Israel and in the U.S. where he has been instrumental in modifying both the technology and the structure of organizations to improve operational effectiveness. He is an honorary member of the American Society for Mechanical Engineers, holds the Sir Leon Bagrit chair in Computer-Aided Design at Ben Gurion University in Beer Sheba, Israel, and is Director of Agile Enterprise Projects at the Agility Forum, as well as Editor-in-Chief of the John Wiley journal, *Agile Enterprise*.

In 1991, Dr. Preiss was selected by the Iacocca Institute of Lehigh University to analyze the U.S. role in the changing structure of worldwide industry. He was one of the facilitators and a co-editor with Steven Goldman and Roger Nagel of the resulting report—*21st Century Manufacturing Enterprise Study: An Industry led View*—which had been commissioned by the U.S. Congress through the Department of Defense.

A dynamic speaker and accomplished writer, his extensive list of published work includes close to 200 original research papers and reports. He co-authored *Agile Competitors and Virtual Organizations: Strategies for Enriching The Customer*, with Steven Goldman and Roger Nagel. He also was chief editor of the *Handbook for Virtual Organization: Tools for Management of Quality, Intellectual Property and Risk & Revenue Sharing*, 1996.

STEVEN L. GOLDMAN is Chief Operating Officer and Chief Technical Officer of the Agility Forum, on loan from Lehigh University where he is an Andrew W. Mellon Distinguished Professor in the Humanities. He has been actively involved in the evolution of agile competition from vision to reality since 1991 when he, Kenneth Preiss and Roger Nagel were the Lehigh co-facilitators and co-editors of the *21st Century Manufacturing Enterprise Strategy* report that launched the agility movement. As an academic, Dr. Goldman holds a joint professorship in Lehigh's Philosophy and History Departments and served for eleven years as Director of Lehigh's Science, Technology and Society Program. His research and teaching address the social relations of modern science and technology. In addition to co-authoring *Agile Competitors and Virtual Organizations* with Roger Nagel and Kenneth Preiss, he has written and/or edited five other books, among them, *Competitiveness and American Society* and *Science, Technology and Social Progress*.

ROGER N. NAGEL is an internationally recognized expert on competitiveness. As Executive Director and CEO of the Iacocca Institute at Lehigh University, Nagel is responsible for 175 people and four corporations dedicated to increasing global competitiveness. The Institute and its associated corporations have worked with several hundred firms across the globe to assist them with competitive strategies, leadership development, and relationship building. The most recent addition to the Iacocca Institute formed under his guidance, is the Agile Learning Center. It is designed to help leaders in creating the future, by developing a vision community and turning vision into action.

While *Business Week*, *Forbes*, *Fortune* and others have described Nagel as being the father of the virtual organization and agile competitor concepts, he credits his co-authors Steve Goldman and Kenneth Preiss and the industrial leaders who worked with him in developing the ideas.

Trained as a computer scientist, Nagel is widely respected for both his technological and business expertise. He is an enthusiastic lecturer who can reach out to and involve a broad spectrum of audiences. His specialty and contribution are in simplifying and communicating new concepts. He has that rare quality of being able to relate to people at many levels, from CEO's to students, and excite them to action. In his role as a communicator, Nagel has worked with a wide variety of groups and industries including advertising, banking, chemical, electronics, food, human resources, manufacturing, pharmaceutical, research, transportation, and utilities.

Nagel has testified at a variety of Congressional hearings. He is a former chairperson and leader of the United States technical delegation in negotiations with Europe, Japan, Australia, and Canada on international collaboration in manufacturing systems. More recently, he was asked by the White House to carry a message to others about President Clinton's desire to share a vision with industry for restoring and maintaining America's international competitive position.

Nagel, who has extensive experience in industry, came to Lehigh University in 1982 from International Harvester. He was named the Harvey Wagner Professor in 1987, and has been honored by several professional societies and international organizations for his accomplishments.